The Anthropocene: Politik—Economics—Society—Science

Volume 21

Series editor

Hans Günter Brauch, Mosbach, Germany

More information about this series at http://www.springer.com/series/15232
http://www.afes-press-books.de/html/APESS.htm
http://www.afes-press-books.de/html/APESS_21.htm

Imtiaz Ahmed
Editor

South Asian Rivers

A Framework for Cooperation

Springer

UNIVERSITY OF DHAKA
ঢাকা বিশ্ববিদ্যালয়

Editor
Imtiaz Ahmed
Department of International Relations
University of Dhaka
Dhaka
Bangladesh

Co-published By:

Bangladesh Environmental Lawyers Association (BELA)
E-mail: bela@bangla.net; Website: belabangla.org

Acknowledgement The cover photograph was taken by Prof. Dr. Imtiaz Ahmed, University of Dhaka who also granted the permission to use it here.

For more on this book, see: http://afes-press-books.de/html/APESS_21.htm

ISSN 2367-4024 ISSN 2367-4032 (electronic)
The Anthropocene: Politik—Economics—Society—Science
ISBN 978-3-319-67373-8 ISBN 978-3-319-67374-5 (eBook)
https://doi.org/10.1007/978-3-319-67374-5

Library of Congress Control Number: 2017952936

Series editor and editor: PD Dr. Hans Günter Brauch, AFES-PRESS e.V., Mosbach, Germany

Printed on acid-free paper

This Springer imprint is published by Springer Nature
The registered company is Springer International Publishing AG
The registered company address is: Gewerbestrasse 11, 6330 Cham, Switzerland

Preface and Acknowledgements

This book is being published at a time when one of the High Courts of India has legally recognized the rivers Ganges and Yamuna as living entities of India. The direction of the Uttarakhand High Court to save the river Ganga "for the generations to come" has revived hopes in the environmentalists who remain deeply concerned by the directions of the Supreme Court of India to implement the disputed River Linking Project "in the national interest" essentially treating shared rivers as national treasures and directing diversion of waters from these rivers on flawed grounds untenable under the prevailing international laws.

This publication includes four articles written by leading experts of the region that, while analysing the existing statist approach and political economy of river management, emphasizes the need for South Asian rivers to be treated as an ecological reality to be governed on common principles of cooperation and legal grounds supported by the collective voices of the people. The introduction and the various chapters identify and analyse the prevailing as well as potential crises and conflicts suggesting needed reforms and basic norms that should guide future treaties/agreements on shared rivers.

While it is expected that the publication shall assist the policy makers of the South Asian countries in rethinking and redoing strategies on transboundary river management, it is also hoped that the analysis of the writers shall add force to the demand of the environmentalists to grant certain rights to our shared rivers that are common heritage of the people of South Asia. If the Parliament of New Zealand can issue legal proclamation declaring the 145-km-long Whanganui River as a living entity, it is perhaps not too much to expect that the common rivers of the region, in their entirety, shall be given the legal status of living entities and that common institutions shall work to protect the rights of these rivers to flow uninterrupted and be saved from pollution.

On behalf of BELA, I would like to thank all the contributors to this book for their thoughtful and analytical work and also for keeping up with the pressing demands for frequent editing. We thank Prof. Imtiaz Ahmed for writing a lucid introduction of what appears to be one of the most contentious issues of the region.

This publication has been made possible with the support from Community Legal Services funded by the UK Government and implemented by Maxwell Stamp PLC, The British Council and the Centre for Effective Dispute Resolution (CEDR). We are grateful to all of them for their generosity!

Syeda Rizwana Hasan
Chief Executive, BELA

Contents

Chapter 1
Introduction

Imtiaz Ahmed

Rivers have always evoked emotions, mainly because they were instrumental in giving rise to life and living, carrying with them, often in folded waves, the narratives of human civilizations. South Asian rivers, small or big, were no exception. So vital were they to life and living to the people of South Asia that the bulk of them ended up being referred to as gods and goddesses, and even to this day are revered by millions of people, indeed, bathing in them with the hope of getting purified if not cleansed of all sins! Some of the South Asian rivers even had a tumultuous birth; at least that is what the myths or retold stories from time immemorial would make us believe. Take the case of River Brahmaputra, named after the "Son of Brahma," whose name is linked to a matricide no less tragic than the Oedipus complex. The story of the river's birth is found in *Kalika Purana* (ca. 10th century). Lord Parashurama, an incarnation of Lord Vishnu, committed a terrible sin by beheading his own mother with an axe allegedly at the instigation of his father. So dreadful was the sin that the axe got stuck in Parashurama's hand. The sages advised him to visit the holy places for salvation. One of the holy places that Parashurama ended up traveling to was Brahmakunda. The place got its name from a child born out of an illicit relationship between Lord Brahma, the creator of the Universe, and Amodha, the beautiful wife of Shantanu, a local sage. The child took the form of water and grew into a great lake, the Brahmakunda, when Shantanu placed him in the middle of four great mountains—Kailash, Gandhamadana, Jarudhi, and Sambwartakk. This is the place where Parashurama arrived with the axe stuck in his hand. He found that the locals of Brahmakunda wanted the release of the waters for cultivation and daily use. Parashurama fulfilled their wishes by cutting down one side of the mountains with his axe. The gods were satisfied and Parashurama to his surprise found the axe separated from his hand. The waters washed off the blood from the axe, leaving a reddish tinge in the river, and that is why it is called the *Luit* in Assamese (from the Sanskrit word for blood). This is how the river was created and came to be called the Brahmaputra, the "Son of Brahma"![1]

[1]Cited in http://www.teachersofindia.org/en/article/story-river-brahmaputra. Accessed on 17 March 2017.

I. Ahmed (ed.), *South Asian Rivers*, The Anthropocene: Politik—Economics—Society—Science 21,
https://doi.org/10.1007/978-3-319-67374-5_1

The mythic birth of the Ganges or Ganga, named after a female deity, is equally noteworthy. The birth of the Ganges is related to supreme devotion and is now considered the most sacred river amongst Hindus. The spiritual justification for this is not difficult to find. The river was first formed in heaven. It is said that Lord Brahma collected the sweat of Lord Vishnu's feet and created Ganga, and since two members of the trinity (Brahma, Vishnu, and Mahesh) contributed to the birth of the river, it became very holy. But that is only the beginning of the story.

King Sagara was fond of horses, but one day one of his sacrificial horses was found stolen. Sagara had 60,000 sons, dull-witted, arrogant, and hot headed, who thought that Kapila, a sage residing in the vicinity of the kingdom, must have stolen the horse.[2] Kapila was in deep meditation when the sons came and accused him of stealing the horse. So angry was the sage, more so for the interruption of his meditation, that he reduced the sons to ashes with his gaze and sent them to the netherworld. Now only the waters of Ganga, then residing in heaven, could bring salvation to the dead sons. It was the great-great-great grandson of King Sagara, King Bhagiratha, who with his supreme devotion, which included rigorous penance for several centuries, finally succeeded in convincing goddess Ganga to descend to earth as a river. But for the descent, which would no doubt be earth-shattering, to take place smoothly, Bhagiratha persuaded Lord Shiva, then residing in the Himalayas, to receive Ganga in the coils of his tangled hair and make the fall softer. And it is because of Shiva's massive tangled hair that the river became divided into several streams, each of which after flowing to different parts of the world finally ended up in the *Ganga Sagar* (or Bay of Bengal, as it is now called), where the ashes of Sagara's sons were purified. They were thereby freed and able to undertake their journey to heaven. Nothing of this would have happened without the supreme devotion of King Bhagiratha, and it is precisely because of his role that the source stream of the Ganges in the Himalayas is named after him to this day.

In this volume, Sudhirendar Sharma and Ajaya Dixit in their respective chapters refer, interestingly, to the *South Asian Water Manifesto*, which was published some two decades back and must have been the region's first water manifesto. The manifesto included a clause emphasizing the sanctity of the rivers:

> Right of the rivers must be codified and guaranteed by the state and the people. Such "rights" have already been codified for oceans and seas. Under the United Nations Convention on the Law of the Sea, it is now the "general duty" of all coastal states to protect and preserve the resources and the riches of the oceans and seas, not simply for the consumption of future generations but for the reproduction of human life itself. River rights can be enacted with similar goals in mind.
>
> In recent times, huge amounts of human and industrial waste are dumped in rivers daily. The polluted rivers not merely spread water-borne diseases, but also raise the cost of treatment of drinking-water. According to one estimate, 114 Indian cities, of at least 50,000 people each, dump raw sewage into the Ganga. The same is true for rivers running through

[2]The story is retold in British Museum, "River Ganga." Available at www.ancientindia.co.uk/staff/resources/background/bg31/bg31pdf.pdf. Accessed on 23 December 2016.

> Kathmandu, Dhaka or many other South Asian cities. *This is an attack on a civilisation defined by its great rivers, where millions of people consider the rivers "holy" and worship them.*
>
> Rivers also have their rights, including the right to be relatively pollution free, to be a safe habitat for riverine forms of life and, within limits, to flow freely. Dams and sewer outlets into rivers interfere with these rights. Large dams and barrages permit the state to restrict and encroach upon the customary rights and practices of local or indigenous peoples and riverine forms of lives. While interference with rivers has contributed to modernisation, bringing electricity and irrigation facilities to a wider section of people, such diversion of water from its natural course has often led to disasters, their magnitude determined by the scale of interference. For example, while large-scale surface irrigation has worsened the problems of water-logging and soil salinity, unregulated withdrawal of water by dams located upstream has dried up downstream beds of several rivers. In places, this has meant an attack on the economic and other life-support systems of the people, in turn leading to the creation of environmental refugees (emphasis mine).[3]

There is no doubt that the authors of the *South Asian Water Manifesto* in the backdrop of water dystopia wanted a deeper appreciation of the civilisational discourse and the sanctity of the rivers recognized and reproduced in contemporary times. The sanctity of the rivers, however, got lost with the territorialisation of the world, a discourse and a mode of action that had origins in the "modern West" with its parallel, but often contradictory, forces of modernity and colonialism playing a formidable but insidious role. South Asia could not help being infected by it, particularly in the wake of physical colonialism and later the colonization of the mind at the hands of the European powers. The Westphalian state became the end and means of all thinking and ended up being a reference point to whatever was pursued—politically, economically, or culturally. As V.G. Hedge was quick to point out in Chap. 2: "The emerging international law on river water sharing requires each of these stakeholders, particularly States, to utilize the rivers in an equitable and reasonable manner." Indeed, so statist did the formulation of rivers become that an international river in South Asia came to be declared as a "national river"!

I am, indeed, referring to the Ganges. On 16 October 2008, Swami Swaroopanand Saraswati, a Hindu religious leader and a former freedom fighter, succeeded in convincing the then Indian Prime Minister, Manmohan Singh, who is incidentally a member of the Indian National Congress and a declared secularist, the need for declaring the Ganges as a national river. In fact, Singh referred to the "special place the Ganga had in the hearts and minds of all Indians," and went on to say that "this emotional link needed to be recognized."[4] This was not a case of honouring the demand of a hydrologist but rather giving into the demand of a religious leader on matters related to rivers. Could this be for electoral politics and

[3]Imtiaz Ahmed, Ajaya Dixit, and Ashis Nandy, *Water, Power and People: A South Asian Manifesto on the Politics and Knowledge of Water* (Colombo: Regional Centre for Strategic Studies, 1997).

[4]Aarti Dhar, "Ganga to be declared a national river," *The Hindu*, 5 November 2008. Available at http://www.Thehindu.com/todays-paper/tp-national/Ganga-to-be-declared-a-national-river/article15335757.ece. Accessed on 23 December 2016.

the desperation of the Congress to win elections in the face of Narendra Modi's *Hindutva* and the fundamentalist wave? But then, if the Ganges is declared as the "national river" of India in recognition to the "emotional link" of the Indians residing there, what about the "emotional link" of the Hindus residing outside India, indeed, in the upper riparian state of Nepal and the lower riparian state of Bangladesh? Would they not be deprived of such "emotional link" as citizens of other states, although the Ganges flows through them as well? It is worth pointing out here that Swami Swaroopanand is an ardent critic of all hydro projects, dams, and barrages on the "holy Ganga". In fact, he urged the Uttarakhand government to go for "alternative power generation measures so as to ensure that the sanctity and flow of the Ganga do not get hindered."[5]

But isn't this what Nepal and Bangladesh, albeit for different reasons, have been demanding from India all along? Could this be then the real source of uneasiness, that Swami's demand had both intra-state and trans-boundary implications, which made the Indian government acquiesce to the demand of the religious leader and declare the Ganges as a national river? Whatever may have been the domestic political compulsions the fact remains that with the declaration and the setting up of the National Ganga River Basin Authority on 4 November 2008, which incidentally would be chaired by the Indian Prime Minister and have as its members only the Chief Ministers of the Indian States through which the river flows, the Ganges—an international river—became fully territorialized and statist! The formidable power of the state, even with respect to the rivers, remains clear. This is where the Westphalian "modern state" differs from the post-colonial state in South Asia. This difference was highlighted by Hamza Alavi some four decades back in what later came to be called as the post-colonial theory of the "over-developed state."[6] Much of the territorialisation of the rivers relates to the latter. Let me explain.

Analytically, the spheres of "modern state" could be divided into three. The first sphere is *political society*, which includes the actors and agencies possessing coercive power. Government, military, police, rules and regulations, all would come under the sphere of political society. Moreover, *triaspolitica* or the "separation of powers," which divides government responsibilities into distinct branches —executive, legislature and judiciary—to prevent the concentration of power and provide for checks and balances, came to be recognized as central to the functioning of political society. The second sphere is the *market* or the one with economic power. The entire vocation of business or moneymaking falls within this domain. In certain circumstances, this sphere could possess enormous power, indeed, to a point where it could become the determining factor in reproducing the state. The third and last sphere is *civil society* or the one that is engaged in the business of reproducing

[5]"Hydropower projects on Ganga not acceptable: Shankaracharya; Go for alternative generation measures, jagad guru tells CM Sandeep Rawat," *Tribune News Service*, Haridwar, 11 June 2012. Available at http://www.tribuneindia.com/2012/20120612/dun.htm#1. Accessed on 23 December 2016.

[6]Hamza Alavi, "State in Post-colonial Societies: Pakistan and Bangladesh," *New Left Review*, 1/74, July–August 1972.

consent or social capital.[7] This would include the academics, students, intellectuals, civil rights groups, political parties, media, cultural bodies, sports clubs and other associations, in fact, all those actors and agencies engaged in reproducing social capital in the society. Each of the domains remains related but "relatively autonomous" from one another.

The post-colonial state in South Asia is markedly different from this. In place of the Westphalian "separation of powers" and "relative autonomy" of the different spheres of the state, the political society in post-colonial South Asia remains "over-developed," indeed, with paramount power to influence and steer the other spheres of the state. Dixit provides an interesting instance of this in Chap. 5:

> With the colonization of the Sub-Continent, the rivers of the GRB [Ganga River Basin]… acquired a political dimension, and their water began to be used to meet political objectives.… In South Asian history, …the political objective became dominant once the British East India Company and later the British state began to develop irrigation systems. The aim was to generate revenue: if irrigation canals could provide sufficient water to increase agricultural production, more tax could be collected.

Economic compulsions or profits otherwise dictated the flow of the river. The colonial legacy, despite protracted anti-colonial struggles, remains alive to this day. The demise of physical colonialism unfortunately made no difference to the colonization of the mind.

Take the case of the constitutional power of post-colonial state on issues related to river water sharing. Indeed, in contrast to the U.S. Constitution and the relative autonomy of the U.S. Supreme Court, which has the power to deliberate and decide on the equitable apportionment of river water, the Constitutions of post-colonial South Asian countries provide no such power to their respective Supreme Courts. As Hedge points out in Chap. 2 with respect to India:

> Article 262 empowers [Indian] Parliament to enact a law that could help in the adjudication of any dispute or complaint with respect to the use, distribution or control of the waters of or in any inter-State river or river valley. Further, it seeks to exclude the jurisdiction of the Supreme Court or any other court in respect of such dispute or complaint.

Only 'experts' commissioned by the government could take up the task of negotiating water conflicts amongst the Indian States. The same is the case with Pakistan and Bangladesh. Any suggestion that India or some other South Asian countries could take lessons from the U.S. Supreme Court to resolve their water-sharing problems, as Hedge seems to advocate, is a non-starter, if not a contradiction, given the qualitative difference in the nature of the state between the U.S. and the countries of South Asia. Such difference, however, cannot be overstretched, as the post-colonial state in many other areas remains structurally organized to reproduce itself in the image of the "modern West," detrimental though it may be to its land and people. An example or two will suffice.

---

[7] Sanjeev Prakash and Per Selle, ed., *Investigating Social Capital: Comparative Perspectives on Civil Society, Participation and Governance* (New Delhi: Sage Publications, 2004).

Despite all the great rivers and having them sanctified by the gods and goddesses, India is in dire need of freshwater. According to one calculation, India has 16% of the world's population but only 4% of its freshwater resources. The per capita availability of freshwater in India has dropped from an acceptable 5,177 m^3 in 1951 to an alarming 1,820 m^3 in 2001. It is estimated that freshwater availability in India could further decline to 1,341 m^3 by 2025, and to 1,140 m^3 by 2050, which is alarming given that the threshold per capita value for water stress is 1,000 m^3.[8] Moreover, when it comes to the availability of water, India ranks 133rd in a list of 180 countries and is significantly below its neighbours—Bangladesh, Sri Lanka, Nepal, and Pakistan, which occupy 40th, 64th, 78th and 80th positions, respectively.[9] Why is India in this situation? Vandana Shiva provides a credible response:

> The deepening water crisis is a result of having treated water as an externality. It was an externality of the Green Revolution, which led to a tenfold increase in water use for the same crop production, thus decreasing water-use efficiency by a factor of ten. This water waste was never part of the definition of productivity and efficiency of industrial agriculture. India's water crisis is rooted in neglecting India's ecology, hydrology, and water heritage. India's water crisis can only be addressed by rejuvenating our water culture and creating water democracy. It is not the stock markets that will decide whether Indian civilization lives or dies. The future of this ancient civilization will be determined by whether our rivers and water systems live or die.[10]

Not so different were the results of constructing large dams, which, apart from storing water for electricity generation irrigation, human consumption, industrial uses and aquaculture, as Dixit aptly pointed out in Chap. 5, ended up displacing people, damaging environmental resources and putting into question social and economic development of the country. *Developmentality* is certainly at fault. As Dixit contends:

> The conventional approach to water, which began in the post-industrialized era and was consolidated in the 20th century water development paradigm, is founded on the erroneous notion that environmental resources such as water are to be consumed or put to work only for the benefit of humans and that any water which flows the sea unused is wasted…. This conception of "waste" ignores the fact that rivers within the hydrological system link oceans and landscape continuously, to which the idea of sink-and-source such as carbon cycle does not apply. Unfortunately, this notion of flowing river as a waste continues even today in developing countries.

Developmentality, however, has wider implications, consuming not only South Asia but also the world, although it must be quickly added that South Asia already had the misfortune of experiencing the worst consequences of it, particularly when it comes to human-led activities and climate change. Earthquakes, floods, cyclones, tsunamis, the melting of glaciers, and the like have consumed South Asia so much that the number of affected people are in the millions, not to mention of the colossal

---

[8]Cited in Binayak Ray, *Water: The Looming Crisis in India* (New York: Lexington Books, 2008), p. xi.

[9]Ibid.

[10]Ibid, pp. xi–xii.

loss of resources. There is, indeed, some truth in the assertion that humans are now a geological force, imbibed with the power of "changing the most basic physical processes of the earth."[11] As Naomi Oreskes maintains:

> For centuries, scientists thought that earth processes were so large and powerful that nothing we could do could change them. This was a basic tenet of geological science: that human chronologies were insignificant compared with the vastness of geological time; that human activities were insignificant compared with the force of geological processes. And once they were. But no more. There are now so many of us cutting down so many trees and burning so many billions of tons of fossil fuels that we have indeed become geological agents. We have changed the chemistry of our atmosphere, causing sea level to rise, ice to melt, and climate to change. There is no reason to think otherwise.[12]

This led some to contend that the earlier geological epoch of Holocene (Whole or Entirely Recent and dated 9,700 BCE) has been replaced by the Anthropocene, with humans becoming a force of nature, tragically contributing to the extinction of species, indeed, as one critic remarked, "similar in intensity to the event around 65 million years ago which wiped out the dinosaurs."[13] Not surprisingly, Dixit is worried, as he cautions us in Chap. 5: "Any loss of Nepal's ice reserves would be catastrophic for the inhabitants of Ganga River Basin, home to some of the world's poorest and most vulnerable people." Can the Anthropocene South Asians rethink developmentality and cooperate on trans-boundary rivers to contain the catastrophe?

This is the moot question raised in the volume. The authors of this volume share a common goal of promoting cooperation amongst the South Asian countries on trans-boundary rivers but the commonality is not limited to promoting cooperation alone. If Hedge takes a statist-legalist position in Chap. 2, advocating a structural-realist position, then Chap. 3, authored by Shafqat Kakakhel, is different only from the standpoint of citizenship. Hedge is an Indian while Kakakhel is a Pakistani. Both intuitively, however, rely on the power of the state and the structures reproducing it. In fact, theoretically their positions are copy images of one another, performing the role of what Antonio Gramsci would call, "organic intellectuals."[14] Sharma and Dixit in Chaps. 4 and 5 respectively, although adhering to an identical goal of achieving South Asian cooperation, remain markedly different from the positions of Hedge and Kakakhel. I will have more to say about this

---

[11]Naomi Oreskes, "The Scientific Consensus on Climate Change: How Do We Know We're Not Wrong?" in Joseph F.C. Dimento and Pamela Doughman, eds., *Climate Change: What It Means for Us, Our Children, and Our Grandchildren* (Cambridge, MA, 2007), p. 93.

[12]Ibid.

[13]Will Steffen, "Humans Creating New 'Geological Age," *The Age*, 31 March 2008. Available at http://www.Theage.com.au/news/environment/humans-creating-a-new-geological-age/2008/03/31/1206850741327.html. Accessed on December 23, 2016.Cited from Dipesh Chakrabarty, "The Climate of History: Four Theses," *Critical Inquiry*, Volume 35, Winter 2009, p. 207. See also Paul J. Crutzen and Eugene F. Stoermer, "The Anthropocene," *IGBP [International Geosphere-Biosphere Programme] Newsletter*, 41, 2000, p. 17.

[14]Antonio Gramsci, *Selections from the Prison Notebooks* (New York: International Publishers, 1971).

shortly, but let me return to the role of "organic intellectuals" and the limits they pose in organizing and reproducing South Asian cooperation, particularly on trans-boundary rivers.

Gramsci once commented, "All men [and women] are intellectuals, but not all men [and women] have in society the function of intellectuals."[15] He then went on to say that "organic intellectuals" are those who perform a "social function" and are *organically* related to a social group, particularly the dominant social group, helping the latter to reproduce its hegemony over the rest of society. But then "organic intellectuals," given their political and economic links to the dominant social group and correspondingly to the interests of the state, are structurally, if not intellectually, ill-placed to resolve conflicts arising from matters that are intrinsically *post-national* or *post-territorial* in nature, as would be the case with trans-boundary rivers. Indeed, if Hedge finds merit in the gradual evolution of the international water discourse, from the Harmon Doctrine of 1896 to the 1997 UN Convention on International Watercourses, then Kakakhel is not far from Hedge in declaring that "South Asian states need to revisit their reservations concerning the UN Watercourse Convention." Both have identical faith in the state, particularly in the ability of the latter to 'revisit' and correct itself without compromising or giving up its socio-politically-constructed interests. This is precisely the reason why the recommendations put forward by Hedge and Kakakhel in Chaps. 2 and 3 respectively, despite holding merit, remain inadequate in institutionalizing the task of water sharing from the standpoint of South Asian rivers. Noteworthy here is the fact that Bangladesh, despite 'sponsoring' and 'favouring' the 1997 UN Convention on International Watercourses when it was endorsed, is yet to ratify the Convention. Since India had abstained from voting the Convention and remains practically opposed to it, many take the non-ratification of Bangladesh as a strategic move on its part not to alienate and annoy India, lest the latter becomes disinterested to negotiate with Bangladesh to resolve their disagreements over sharing river water. Politics otherwise dictates legalism and the nature of cooperation in South Asia!

Sharma and Dixit, on the other hand, in their respective chapters make a departure from statist positions and flag a *post-statist* or *post-territorial* understanding of sharing river water in South Asia. Incidentally, both of them are not so much led by the power of the state, particularly when it comes to sharing river water, knowing well that the downstream flow cannot be stopped, nor would the terrain, for instance, justify the harnessing of the hydropower potential by Nepal alone. Indeed, if it were to become economically viable, neighbouring countries, including India, ought to contribute to the development of Nepal's hydropower potential. Post-territoriality is otherwise more meaningful to Nepal, particularly when it comes to harnessing river water, and this prompts Sharma and Dixit to question and rethink the state of water knowledge in South Asia. Both emphasize different things, however.

---

[15]Ibid.

Admittedly, there is an element of Indo-centrism in Sharma, similar to Kakakhel, when critiquing the dismal state of riverine cooperation in South Asia. As Sharma contends in Chap. 4, "Unless India comes good on its water cooperation agenda, the foundation of regional cooperation will continue to remain weak at its core." Sharma is quick to suggest "people-to-people negotiations on co-operative management of shared river basins." But this will require a "new conceptual order," as Sharma reasons, and following Tagore suggests making use of South Asia's civilizational strength, which incidentally lies more with its 'cultures' than in politics. In fact, Sharma's optimism with South Asia's cultural heritage has no bounds and remains an eye-opener:

> Multiple cultural maps of South Asia, superimposed on the river basins, can bind communities with common cultural threads. While a basin map capturing the common Hindi chord along the trans-boundary rivers could infuse new hydro-solidarity amongst riparian communities in Nepal and India, a sub-basin map contoured around the Bangla speaking populace could easily build a better appreciation for upstream-downstream concerns between riverine communities of India and Bangladesh.

Dixit, on the other hand, is more focused on the knowledge-system, claiming that a transformation of the existing epistemology of water could make a difference to our understanding of rivers and help transform the inadequate, if not distorted, understanding of water. As Dixit notes in Chap. 5:

> A different educational context and a different decision-making approach are missing in the trans-boundary water policy terrain in the GRB and in South Asia as a whole. Creating these two things could be a starting point for conceiving of a new compact for stewarding South Asian water.

But this would require a "creative engagement" with water, indeed, with the objective of nurturing not only an "inter-disciplinary" water discourse but also a newer epistemology of water. Anything short of this would limit the goal of institutionalizing river water sharing amongst the South Asian countries.

Does this mean that Sharma and Dixit stand in opposition to the perspectives put forward by Hedge and Kakakhel? In some ways, yes, and that is where the merit of this volume lies. If the latter two cater to a *statist* position, Sharma and Dixit make best of *post-statist* understanding of things. This brings me back to where I started. In the beginning I have referred to a narrative filled with contradictions, indeed, with respect to the birth of both Brahmaputra and Ganges, rooted as they were, as I would contend, to an indigenous form of dialectics called *prasangika*. The latter stands for the "method of examining all possible alternative interpretations of the opponent's proposition, showing the absurdity of the respective consequences and thus refuting it."[16] In many ways, that is precisely what Sharma and Dixit

[16]Esther Solomon, *Indian Dialectics: Methods of Philosophical Discussion*, Volume II (Ahmedabad: Gujarat Vidya Sabha, 1978), 520. See also, Imtiaz Ahmed, "Diversity in dialectics: a methodological quest for en-gendering security." Paper presented at the International Conference on *Gender, Conflict and Security: Perspectives from South Asia*, organized by South Asian University and UN Women in New Delhi, India, 23–24 April, 2015.

unwittingly do, but then, without Hedge and Kakakhel, their remedies become incomplete, if not half-witted. In this light, one can say with some confidence that this volume is bound to provide a different but instructive reading of the discourse on South Asian trans-boundary rivers and contribute to the current campaign of overcoming the water dystopias in South Asia. Let us keep our hope alive!

Chapter 2
National and International Legal Aspects of River Water Sharing: The South Asian Experience

V.G. Hegde

Conflicts over shared rivers between various stakeholders usually arise because of the depletion of water flows due to diverse factors and circumstances broadly classified as (a) geographic, hydrographic, hydrological, climatic, ecological and other factors of natural character; (b) the socio-economic needs of the population; (c) the quantum of population dependent upon the river; (d) the effects of the use or uses of the river by one stakeholder on another; (e) existing and potential uses of the river; (f) conservation, protection, development and economy of use of the river waters and the costs of measures taken to that effect; and (g) availability of alternatives, of comparable value, to a planned or existing use of the river.[1] The emerging international law on river water sharing requires all stakeholders, particularly States, to utilize the rivers in an equitable and reasonable manner. The rivers should be used and developed with a view to attain optimal and sustainable utilization taking into account the concerns of all stakeholders.[2]

The per capita use of fresh water resources is increasing at a phenomenal pace within the global context on account of the rise in global population, migration to

---

[1]For the categorization of these factors and circumstances, see Article 6 of the *United Nations Convention on the Law of the Non-Navigational Uses of International Watercourses,* United Nations General Assembly, A/RES/51/229, July 8, 1997.

[2]Ibid. See Article 5, which imposes on the States (as stakeholders) not only to develop and protect the rivers, but also requires them to cooperate with other States in the protection and development of the rivers.

I. Ahmed (ed.), *South Asian Rivers*, The Anthropocene: Politik—Economics—Society—Science 21,
https://doi.org/10.1007/978-3-319-67374-5_2

cities, and other related factors.[3] In certain developing countries, the extensive use of river water for agricultural and industrial production has resulted in the depletion of the flow of water during lean seasons. The flow of pollutants to rivers and other similar factors have given rise to conflicts among stakeholders who might be located within different jurisdictions.

Disputes over the sharing of river water have also arisen within the domestic context of States. The different federal units of a State through which a river flows, for example, may have differences and strong contentions with regard to the sharing of the river water. The various norms and principles developed several decades ago by the various contending federal states within the United States, for example, with regard to sharing of river water resources are notable and have contributed to the evolution of international law on this issue as well.[4] These contentions within the United States with regard to the sharing of river waters among its federal units went on for a long time and internally the US Supreme Court had to deal with these issues from time to time. Similar examples exist in different parts of the world.

In this context, the river water sharing issues within South Asia are no different. The three main rivers, Ganges, Brahmaputra, and the Indus largely form the core of the river water sharing challenge in South Asia.[5] Bangladesh, India, Nepal, and

---

[3]It is estimated that of all the water on earth, only 2.5% is fresh water and 1% is easily available for human use. Despite the complexity of the problems, the last 50 years have seen only 37 acute disputes involving violence, compared to 150 treaties that have been signed. The focus of negotiation and treaty-making in the last century has shifted away from navigation towards the use, development, protection and conservation of water resources. For details on Transboundary Water Issues in the context of International Decade for Action "Water for Life" 2005–2015 and other relevant articles and literature, see www.un.Org/waterforlifedecade/transboundary_waters.shtml. Accessed on 20 December 2015. Also see Food and Agriculture Organization of the United Nations, "Land and Water." Available at http://www.fao.org/land-water/en/. Accessed on 20 December 2015.

[4]Some of the major cases decided by the US Supreme Court are *Kansas v. Colorado* (1907)—this was about diversions from Arkansas River in Colorado that deprived Kansas of water; *Wyomingv. Colorado* (1922) wherein Wyoming brought proceedings against Colorado for diverting water from the Laramie River; *Connecticut v. Massachusetts* (1931) this was about the diversion of the waters of the Connecticut River, as well as the Ware and Swift Rivers, by Massachusetts; *New Jersey v. New York* (1931) about New York diverting waters of the Delaware River and its tributaries; *Nebraska v. Wyoming* (1945) about the equitable apportionment of the North Platte River. For an account of brief discussion on these cases see Stephen C. McCaffrey, *The Law of International Watercourses*, (Oxford University Press: New York, 2001) pp. 221–228. A brief summary of some of these cases has been incorporated in the next section on the Indian Context.

[5]The Ganges rises in the Himalayas and flows through India and Bangladesh. In Bangladesh it joins the Brahmaputra and is called as Padma. Later, it empties into Bay of Bengal forming a huge delta area. The Ganges and Brahmaputra-Meghna (GBM) drainage basin has a combined area of about 1,600,000 km^2. This combined basin area spreads across Bangladesh, Bhutan, India, Nepal and China. The Indus rises in Tibet and flows through India and Pakistan. It has a length of about 3,200 km. See Stephen C. McCaffrey, *The Law of International Watercourses*, (Oxford University Press: New York,, 2001) p. 248; also see B.S. Chimni, "A Tale of Two Treaties: The Ganga and Mahakali Agreements and the Watercourse Convention" in Surya P. Subedi, (Ed.) *International Watercourses Law for the 21st Century: The case of the River Ganges Basin* (Ashgate Publishing Limited: Oxon, 2005) p. 64.

Pakistan on account of their geographical proximity and the historical attachment of their populations to the existing river systems pose complex water sharing issues within South Asia. Besides these three river systems, there are other smaller rivers that also form part of the entire river water sharing issue within South Asia.

In terms of South Asian transboundary rivers, India is in a peculiar position. For example, it has upper riparian status vis-à-vis Pakistan in sharing the Indus and its tributaries and vis-à-vis Bangladesh in sharing the waters of Ganges, Brahmaputra, and several other smaller rivulets including Teesta. On the other hand, it has lower riparian status vis-à-vis Nepal and China. With this dual status as a lower and upper riparian country, India is in a delicate position in terms of developing balanced and viable legal and technical arguments to justify its need for augmenting water usage.

The partition of the South Asian sub-continent has essentially given rise to all these contentious transboundary river water sharing issues.[6] Otherwise, for a very long time, a large population has generally accessed and been dependent upon these rivers for their needs.[7] This long-term dependency on these rivers by the people in these countries has not only had socio-economic implications, but also given rise to cultural, civilisational and spiritual affinities.[8] In this sense, for the people of the sub-continent, these rivers are part of their life and existence.

Notably, these three major transboundary rivers have a gigantic basin area that includes several of its tributaries as well. After a protracted and difficult negotiation of nearly ten years, both India and Pakistan succeeded in concluding the Indus Water Treaty of 1960, which now governs the water sharing arrangement between India and Pakistan for an indefinite period.[9] In contrast, the sharing arrangements and agreements for the waters of the Ganges between India and Nepal generally had a short-term approach till the conclusion of Mahakali Treaty in 1996. This Treaty ended the short-term approach and sought to govern the integrated development of the Mahakali River for a period of seventy-five years with a ten-year review

---

[6]For an historical account of these issues, specifically with regard to the Indus see N.D Gulati, *IndusWaterTreaty: An Exercise in International Mediation* (Allied Publishers: Bombay,1973); and A.A. Michel, *The Indus Rivers: A Study of the Effects of Partition* (Yale University Press: New Haven, 1967).

[7]The population dependent on the entire Ganges basin is approximately 300 million (of which 10 million are in Nepal and 40 million in Bangladesh and the rest in India). Similarly, the Indus has a drainage basin of approximately 1,165,000 km^2. The network canals connect a large area of the basin in Punjab area (both within India and Pakistan); see McCaffrey, n. 2, p. 248.

[8]It has been pointed out "The law of international watercourses has developed in tandem with the evolution of human social organization and the intensification of use by human societies of freshwater…It is well known that rivers nourished the great ancient civilizations and drove their economies…But even before the rise of these civilizations evidence of early canals and dikes suggests that small communities had found it necessary to cooperate in order to control and utilize effectively of major rivers". See McCaffrey, n. 1. p. 58.

[9]For the text of the treaty see http://wrmin.nic.in/writereaddata/Inter-StateWaterDisputes/Volume-I1920752696.pdf and also http://www.worldbank.org/en/region/sar/brief/fact-sheet-the-indus-waters-treaty-1960-and-the-world-bank. Accessed on 5 January 2013.

clause.[10] Similarly, both India and Bangladesh discarded their short-term approach with the signing of the Farakka Treaty in 1996, which has a validity of thirty years.[11] The sharing of the waters of the Teesta River with Bangladesh is also on the anvil. Due to various internal domestic political compulsions of India the signing of this treaty has been deferred.[12]

In the next section, this paper, considering the above complexities in the sharing of river water, proposes to examine the domestic constitutional and normative schemes that exist within the various South Asian States for the resolution of river water sharing issues. The section thereafter will review various international legal norms as embodied in bilateral and multilateral agreements and arrangements that seek to regulate river water sharing mechanisms. The primary emphasis, however, will be on South Asia and the countries within South Asia.

2.1 River Water Sharing in South Asia: Domestic Legal and Policy Framework

States usually provide for domestic legal or policy framework to share freshwater resources that exists or available within their territory. There are several layers of legal and regulatory framework that are in place to protect these natural resources. As we could see in the ensuing discussion some States attempt to do this in more general terms and others with specific provisions to deal with any river

[10]For the text of the treaty see http://internationalwaterlaw.org/documents/. Accessed on 5 January 2013.

[11]For the text of the treaty see http://africanwater.org/farakka_water_treaty.htm. Accessed on 5 January 2013.

[12]In March 2010, at their 37th Ministerial level Joint River Commission meeting, India and Bangladesh finalised the decision to sign the Teesta River Agreement. Drafts were exchanged between the two countries and the agreement was supposed to be signed in September 2011 during the visit of the Indian Prime Minister Manmohan Singh. That did not happen due to domestic political reasons within India. The Teesta River enters Bangladesh near Nilphamari district and flows 45 km through the agri-dominated districts of Rangpur, Lalmonirhat, and Gaibandha before meeting the Brahmaputra River in Kurigram. The Teesta River barrage at Gozaldoba in India regulates the amount of water flow downstream to Bangladesh. In order to increase the irrigation potential of the northwest region, Bangladesh constructed the Dalia barrage on the Teesta River in Lalmonirhat district to provide irrigation water from the river. The Teesta river flows from Sikkim, and cascades through North Bengal, before entering Bangladesh. The West Bengal Government had agreed on sharing of up to 25,000 cusecs of water, but the final version of the agreement aimed at sharing 33,000–50,000 cusecs, which would hurt the interests of West Bengal. The other argument is that the Teesta's lean season water flow has not been jointly studied and surveyed by Indian and Bangladeshi experts. Some argue that the surface and ground potential of water resources in the Brahmaputra Basin, consisting of sub-basins of the Sankosh, the Raidak, the Torsa, the Jaldhaka, and the Teesta rivers are apparently sufficient to meet the requirements of West Bengal and Bangladesh.

water-related conflicts.[13] The water-related rights within the domestic context of any State need to be categorized into two different frameworks. The first categorization is in terms of specific water rights as regulated by the State which could include such rights as access to clean drinking water, water usage and rights that restrict or vest those rights in certain sections or groups of people, environmental issues, and other related rights. These rights are enshrined in specific domestic legislations such as India's Water Act, Irrigation Act, and related enactments (essentially all acts that relate to water use management). The second categorization comprises rights and policies regarding the actual sharing of rivers and other freshwater resources such as lakes and ponds within the domestic context. This is an issue that could arise within and between federated units or provinces of a State. The river-sharing legal frameworks within these federal units, as mentioned above in the context of United States for example, contributed to the development of international legal norms as well. These issues are examined in the latter part of the paper.

2.2 Protecting Water Resources: The South Asian Constitutional Context

The Constitutions of the South Asian countries do not directly refer to the protection and preservation of water resources although some of them seek to deal separately with the resolution of river water disputes within their boundaries.[14] The Indian Constitution, for instance, initially had no provision relating to the protection and preservation of environment and forests. In 1976, it inserted an amendment to include Article 48-A, which provided that "The State shall endeavour to protect and improve the environment and to safeguard the forests and wild life of the

[13]Generally on the domestic legal frameworks within the South Asian context see R. Ramaswamy Iyer (Ed.), *Water and Laws in India*, (New Delhi: Sage Publications, 2009); Tushar Shah, *Taming the Anarchy: Groundwater Governance in South Asia*, (New Delhi: Routledge, 2009); Kishor Uprety and Salman M. A. Salman, "Legal Aspects of Sharing and Management of Transboundary Waters in South Asia: Preventing Conflicts and Promoting Cooperation", (2011), *Hydrological Sciences Journal*, Vol. 56 (4) pp. 641–661; Gopal Siwakoti, "Transboundary River Basins in South Asia: Options for Conflict Resolution". Available at https://www.internationalrivers.org/sites/default/files/attached-files/transboundaryriverbasins.pdf. Accessed on December 20, 2015; Shanta Mohan, Sailen Routray, and N. Shashikumar (Eds.), *River Water Sharing: Transboundary Conflict and Cooperation in India*, (New Delhi: Routledge, 2010).

[14]Both India and Pakistan provide separate provisions for conflict resolution for such disputes by excluding any jurisdiction for the Courts. Article 262 of the Indian Constitution seeks to provide for a separate law to adjudicate "any dispute or complaint with respect to the use, distribution or control of the waters of, or in, any inter-State or river valley." Pakistan provides in Article 155 of its Constitution that any complaint with respect to the use and distribution or control of water by the Federal Government or the Provincial Government should be handled by a Council in which Prime Minister and Chief Ministers of all the Provinces are members. It also excludes jurisdiction of any Court in these matters.

country."[15] Similarly, Bangladesh inserted an amendment to its Constitution in 2011 to include Article 18A which states that, "The State shall endeavour to protect and improve the environment and to preserve and safeguard the natural resources, biodiversity, wetlands, forests and wildlife for the present and future citizens." There are similarities in these obligations created within the Indian and Bangladeshi Constitutions both in terms of the placement of these provisions and their wider interpretations to include water bodies and rivers.

The Sri Lankan Constitution makes a reference in its Article 27(14) to "protect and preserve and improve the environment for the benefit of the country". However, Sri Lanka has nearly fifty legislations that deal with water resources and related issues and also about forty agencies dealing with these issues.[16] The new Nepalese Constitution in Article 51(g) outlines its policy on various aspects of preservation and management of environment and natural resources.[17] Nepal, while pursuing policy of "conserving the natural resources and imbibing the norms of inter-generation judicious use of it for the national interest" seeks to achieve a fair distribution of benefits by giving local people priority and preferential rights. Specifically, Nepal provides in Article 51(g)(2) that "The State shall pursue a policy of prioritizing national investment in water resources based on people's participation and making a multi-utility development of water resources." Bhutan, in its Constitution, has several provisions on preservation of environment, its culture and traditions, and related issues. Article 5 specifically deals with the 'environment' and regards every Bhutanese as a trustee of the Kingdom's natural resources and environment for the benefit of present and future generations. There is also reference to the State's ability to extend special protection to any part of Bhutan by declaring it a "nature reserve", "critical watershed, or such other categories meriting protection".[18]

A brief survey of all the South Asian Constitutions shows that there are general references to the protection and management of environment and related aspects. The legal and policy framework relating to the regulation of water related issues,

[15]Article 48-A is in Part IV (Directive Principles of State Policy) of the Constitution of India, which is unlike Part III (Fundamental Rights) is non-justiciable. Article 37 of the Indian Constitution provides that the provisions contained in Part IV are not to be enforceable by any court, "but principles therein laid down are nevertheless fundamental in the governance of the country and it shall be the duty of the State to apply these principles in making laws".

[16]These agencies include both governmental and community-based agencies. Governmental agencies usually undertake regulatory functions, the community-based agencies seek to supplement and implement these regulatory and policy framework at the grassroots. For detailed account on this see V. K. Nanayakkara, "Sri Lanka's Water Policy: Themes and Uses". Available at http://publications.iwmi.org/pdf/H042809.pdf. Accessed on December 20, 2015.

[17]Article 51 of the Nepalese Constitution is titled as "State Policies" and deals with various national policy aspects relating to important areas, which *interalia*, also include in clause (g) policies regarding the conservation, management, and use of natural resources.

[18]Article 5(5) of the Bhutanese Constitution provides "Parliament may, by law, declare any part of the country to be a National Park, Wild Life Reserve, Nature Reserve, Protected Forest, Biosphere Reserve, Critical Watershed and such other categories meriting protection."

including the sharing of river waters flow from these broad constitutional provisions.[19] As elaborated in the following discussion, both India and Pakistan seek to exclude any intra-State conflict or dispute from the purview of formal adjudicatory bodies like courts, including their Supreme Courts. Any such conflict, as elaborated in their respective Constitutions will have to be resolved through the establishment of Tribunals or Commissions which will be comprised of water engineers and legal experts.

In this regard, therefore, both these countries follow the United States model in resolving intra-State water conflicts through the formation of expert bodies that would take into account concerns of all stakeholders and provide a solution based on the practical assessment of the ground reality. In adjudicatory process this kind of flexibility is not available to the stakeholders resulting in the non-acceptability of the final verdict as it becomes difficult for them to convince their affected population. It should also be noted that intra-State conflicts, like the transboundary conflicts, also involve dealing with complex socio-economic and political dimensions. It might become difficult for the stakeholders such as provincial units or federal units to convince members of their populace who are adversely affected by these decisions. Negotiated and mediated decisions, in such a scenario, seem to be the best alternative and some of the South Asian countries have adopted this method while dealing with their intra and inter-state conflicts.

2.3 Legal and Policy Framework

2.3.1 National Policies

Bangladesh, India, Nepal, and Pakistan[20] with greater stakes in river water sharing, both intra and inter-State, have a declared national water policies. This policy framework of these countries emanate, besides their constitutional requirements, from their relevant national legislations such as Water Acts and other related

[19]Afghanistan seems to be the only exception, as it does not have any specific or remotely connected provision on water or environment related issues.

[20]For the text of the National Water Policy of Bangladesh, 1999, which remains part of the Water Act see http://mowr.portal.gov.bd/sites/default/files/files/mowr.portal.gov.bd/files/32e67290_f24e_4407_9381_166357695653/National%20Water%20Policy%20(English).pdf. Accessed on December 20, 2015. For the draft National Water Policy of Pakistan, 2004, see https://www.waterinfo.net.pk/sites/default/files/knowledge/National%20Water%20Policy%20%28Draft%29.PDF; also see for a detailed account on Pakistan Water Policy, Politics and Management, Medha Bisht, *Water Sector in Pakistan: Policy, Politics, Management"* IDSA Monograph Series, No. 18, April 2013. Available at http://www.idsa.in/system/files/Monograph18.pdf. Accessed on 20 December 2015. For the text of the National Water Policy of India, 2012 outlining the various aspects of the water use and management, see http://wrmin.nic.in/writereaddata/NationalWaterPolicy/NWP2012Eng6495132651.pdf. Accessed on December 20, 2015. For the text of the National Water Plan of Nepal, 2002, see http://www.moen.gov.np/pdf_files/national_water_plan.pdf. Accessed on December 20, 2015.

enactments. All these national policies generally deal with aspects relating to water use and management, conservation, water pricing, water user associations, enhancement of water availability, flood and drought, institutional arrangements with particular emphasis on creating legal framework and suitable water laws to deal with intra and inter-state (transboundary) water problems, issues relating to and creation of database and information systems, research and training, implementation, and other related issues.

The need for a national water policy has been aptly summed up in the following observation by the Prime Minister of Bangladesh as it brings "…order and discipline in the exploration, management and use of water resources…to manage the water resources of the country in a comprehensive, integrated and equitable manner…policy…successfully integrated internationally accepted water management principles, norms and standards, with the demanding of social and economic needs of a developing country."[21] Flowing from this national policy, the Bangladesh Water Act was enacted in 2013 to "make provisions for integrated development, management, abstraction, distribution, use, protection and conservation of water resources."

One of the important aspects of the national water polices of all South Asia countries is the creation of what has been termed as a "National Council" or such similar bodies to oversee the entire legal and policy framework. This institutional arrangement is regarded as the highest decision-making body on all policy matters relating to the sharing of water resources and also as instrumental in resolving internal conflicts among federal units. Some of the common objectives could be summed up as (a) making policies and directions for the integrated development and use of national water resources; (b) taking stock of national water resources and planning; (c) developing and approving national resource plans; (d) taking stock of the availability of water resources; and (e) addressing flood and drought management issues.[22]

Other unique feature of these policies is that they note the particular challenges associated with the States' lower riparian status. In the case of Bangladesh and Pakistan both are lower riparian States and therefore, as stated in the policy, water availability is severely restricted by their location. In its national policy, Bangladesh, for example, notes that, "due to its location as the lower-most riparian, Bangladesh has no control over the rivers entering through its borders." Floods and water scarcity, which occur frequently, within Bangladesh in the context of three major basins, namely Ganges, Brahmaputra and Meghna, need careful

[21]Foreword written by the Prime Minister of Bangladesh to the National Water Policy of Bangladesh adopted in 1999 and these words remain valid even to this day. See National Water Policy of Bangladesh, 1999. Available at http://mowr.portal.gov.bd/sites/default/files/files/mowr.portal.gov.bd/files/32e67290_f24e_4407_9381_166357695653/National%20Water%20Policy%20(English).pdf. Accessed on December 20, 2015.
[22]Ibid.

consideration. Therefore, the policy of Bangladesh is to work closely with the co-riparian States to conclude viable treaties and sharing arrangements.[23]

2.3.2 *Institutional Arrangements*

The other notable aspect of the national water policies in the South Asian context is that the institutional arrangements contemplated in these policy frameworks are chaired at the highest level. The advisory councils and panels constituted pursuant to these national policies are generally chaired by the Prime Minister of the respective countries. That itself shows the importance of the water policy framework and its implementation process to these countries. The National Water Resources Council created under the Water Act, 2013 of Bangladesh is chaired by the Prime Minister and all the other major Ministries are part of this Council, besides several other experts and representatives of non-governmental bodies.[24]

A similar institutional framework has been adopted by Pakistan, India, and Nepal in their respective national water policies. Pakistan, for example, has a constitutional body called the "Council of Common Interests" (CCI). The CCI is chaired by the Prime Minister and includes the Chief Ministers of all the Provinces as members.[25] This Council takes up the issue of river water sharing and related disputes that may arise between the provinces. This Council, upon receiving any such complaint relating to water sharing, could either decide the matter on its own or request the President of Pakistan to appoint a Commission consisting of such persons having special knowledge and experience in irrigation, engineering, administration, finance, or law.[26] If the matter is before the Council, even the jurisdiction of any Court is excluded.

[23]Ibid. The National Water Policy of Bangladesh, therefore, contends, "It may take considerable effort and time for Bangladesh to work out joint plans for different river basins with other co-riparian countries. As a long-term measure, therefore, it is the policy of the government to undertake essential steps for realising basin-wide planning for development of the resources of the rivers entering its borders."

[24]According to Section 4 of the Bangladesh Water Act, 2013, the Council, besides the Prime Minister, includes the Ministers of Finance, Agriculture, Planning, Local Government, Rural Development and Cooperatives, Law and Justice, Land Water Resources, Foreign Affairs, Fisheries, Shipping, and Environment and Forests. The range of membership in the Water Resources Council shows the importance attached by the concerned countries of South Asia to the Water Resource-related issues.

[25]Article 153 of the Constitution of Pakistan.

[26]Article 155 of the Constitution of Pakistan outlines the provisions to deal with the complaints relating to water supplies. It, *interalia*, provides, "(1) If the interests of a Province, the Federal Capital, or the Federally Administered Tribal Areas, or any of the inhabitants thereof, in water from any natural source of supply (or *reservoir*—inserted in 2010) have been or are likely to be affected prejudicially by (a) any executive act or legislation taken or passed or proposed to be taken or passed or; (b) the failure of any authority to exercise any of its powers with respect to the use and distribution or control of water from that source; the Federal Government or the Provincial

The sharing of waters of the Indus and other smaller rivers is a contentious issue within Pakistan. It should be noted that Pakistan has one of the largest irrigation canal systems. Immediately after the partition of India, there were several contentious and conflicting issues between both the countries. The Indus Treaty of 1960 could be negotiated only upon the intervention of the World Bank and the Treaty itself has the stamp of the Bank in its working and in particular with regard to its dispute settlement clauses. According to one view the problem of water sharing is a reflection of the deterioration of interprovincial relations within Pakistan.[27]

Despite the creation of the Indus River System Authority (IRSA) and a Water Accord in 1991 the issue of river water sharing has been a contentious issue among Sindh, Punjab, and Khyber Pakhtunkhwa (KPK). In 2002, a Parliamentary Committee was established to resolve these interprovincial differences regarding water sharing and to develop consensus. Despite all these efforts within Pakistan, the differences regarding water sharing persist.

Sindh believes that the waters of the Indus historically and exclusively belong to it. There is also the feeling that Punjab by constructing new projects to connect the tributaries of the Indus is a dominating actor in the entire interprovincial conundrum. Sindh argues that Punjab is transferring Indus waters through two link canals, namely, Chashma Jhelum and Taunsa-Panjnad. Both the Tarbela and Mangala dams which have been built to provide water replacement of three of its eastern rivers, Ravi, Sutlej, and Beas have also become controversial as they are being opposed by Sindh, which is asking for its fair share of water. There are several serious contentions. Sindh claims that Punjab and KPK take away water through

(Footnote 26 continued)

Government concerned may make a complaint in writing to the Council. (2) Upon receiving such complaints, the Council shall, after having considered the matter, either give its decision or request the President to appoint a Commission consisting of such persons having special knowledge and experience in irrigation, engineering, administration, finance, or law as he may think fit, hereinafter referred to as the Commission. (3) Until [the Majlis-e-Shoora (Parliament)] makes provision by law in this behalf, the provisions of Pakistan Commissions of Inquiry Act, 1956, as in force immediately before the commencing day shall apply to the Council or the Commission as if the Council or the Commission were a Commission appointed under that Act to which all the provisions of section 5 thereof applied and upon which the powers contemplated by section 10A thereof had been conferred. (4) After considering the report and supplementary report, if any, of the Commission, the Council shall record its decision on all matters referred to the Commission. (5) Notwithstanding any law to the contrary but subject to the provisions of clause (5) of Article 154, it shall be the duty of the Federal Government and the Provincial Government concerned in the matter in issue to give effect to the decision of the Council faithfully according to terms and tenor. (6) No proceeding shall lie before any Court at the instance of any party to a matter which is or has been in issue before the Council or of any person whatsoever, in respect of a matter which is actually or has been or might or ought to have been a proper subject of complaint to the Council under this Article.

[27]For an account on interprovincial disputes that exist within Pakistan, see Ahmed Hayat Khan, "Water Sharing Dispute in Pakistan: Standpoint of Provinces", *Berkeley Journal of Social Sciences*, vol. 4, Spring 2014; and Medha Bisht, *Supra* note 21.

barrages falling under its territorial jurisdiction. Punjab, on the other hand, points out that there is huge water loss between the Sukkur and Kotri barrages. The Kalabagh dam, a mega structure supported by Punjab, is being opposed by all the other provinces. Sindh, KPK, and Baluchistan Provincial Assemblies have passed resolutions opposing the building of this dam. Pakistan, through its Article 155 of its Constitution, is attempting to resolve these river sharing conflicts by initiating negotiations.

2.3.3 *Exclusion of Formal Domestic Adjudication*

India has a provision in its Constitution, which seeks to exclude the jurisdiction of its highest court of the land in resolving its internal water sharing disputes. There are several keenly contested and on-going disputes among various States of the Indian Union.[28] Article 262 empowers Parliament to enact a law that could help in the adjudication of any dispute or complaint with respect to the use, distribution, or control of the waters of or in any inter-State river or river valley. Further, it seeks to exclude the jurisdiction of the Supreme Court or any other court in respect of such

[28]These pending disputes are—(a) the sharing of the waters of Ravi and Beas between Punjab, Haryana and Rajasthan. Earlier in 1981 an agreement regarding sharing of the waters had been concluded between the Chief Ministers of these States. However, in 2004 Punjab enacted the Punjab Termination of Agreements Act, 2004 terminating all earlier agreements including the 1981 agreement fully discharging Government of Punjab of any obligation arising from the agreements. A Presidential Reference regarding this Act has been pending under Article 143 of the Constitution: (b) the sharing of the waters of Vansadhara river between Orissa and Andhra Pradesh. Orissa made a complaint in February 2006 to the Central Government against Andhra Pradesh for constructing a flood flow canal and thereby affecting the use, distribution and control of waters of the river. The basic contention of the State of Orissa in the complaint is that the flood flow canal would result in drying up of the existing river bed and consequent shifting of the river affecting ground water table; (c) the sharing of the waters of Mandovi River for which a complaint was made by Goa in July 2002 against Maharashtra and Karnataka. Goa is demanding assessment of available utilisable water resources in the basin at various points and allocation of this water to the three basin States keeping in view priority of the use of water within basin as also to decide the machinery to implement the decision of the tribunal; (d) the sharing of the waters of Krishna River between Karnataka, Maharashtra, and Andhra Pradesh. The Krishna Water Dispute Tribunal has already submitted its report and it has been notified by the Government of India in 2010. See http://wrmin.nic.in/writereaddata/Inter-StateWaterDisputes/KWDTReport9718468760.pdf. Accessed on December 20, 2015. The Implementation Board has been functioning; (e) the sharing of the waters of Cauvery between Karnataka and Tamil Nadu. The Cauvery Water Disputes Tribunal was constituted in June 1990 itself and it had submitted its final report in 2007. At present, monitoring of the implementation of the orders of Cauvery Tribunals is in question under the Cauvery River Authority and Cauvery Monitoring Authority. For details about the recent status of these cases see http://wrmin.nic.in/forms/list.aspx?lid=384&Id=4. Accessed on December 20, 2015.

dispute or complaint.[29] Pursuant to this provision, the Indian Parliament enacted the Inter-State River Water Disputes Act, 1956. This law provides for the creation of tribunals to settle inter-State river water disputes upon a complaint by a State that has been adversely affected. Before taking recourse to dispute settlement mechanisms under this enactment States must first hold negotiations to resolve the dispute. Only after the failure of such negotiations can a dispute be entertained under the 1956 Act. The Central Government has an obligation to create a tribunal within a specific timeframe to resolve the matter. These tribunals are constituted with expert members who will have to decide the sharing formula based on the technical arguments put forward by the States.[30]

India now has more than five decades of experience in dealing with its internal inter-State disputes relating to water sharing. In the South Asian context, India is the only country that has attempted to solve some of these difficult disputes between its States. These disputes were formally submitted for resolution, as per its Constitutional mandate under Article 262, to Tribunals. This process preceded a long period of failed negotiations and in some cases contentions between the populations of the States were too emotional, resulting in huge law and order problems as well. Some of these Tribunals had to look for an applicable law and principles to arrive at rational distribution criteria. What are these principles and criteria? In the following discussion we will examine some of these aspects of international legal principles as used by the domestic tribunals of India to resolve river water disputes.

The Narmada and Godavari tribunals were the earliest ones to be constituted and the main issue they sought to resolve concerned the equitable apportionment of available water in the rivers among contending States.[31]

---

[29]Article 262 provides "(1) Parliament may by law provide for the adjudication of any dispute or complaint with respect to the use, distribution or control of the waters of, or in, any inter-state river or river valley. (2) Notwithstanding anything in this Constitution, Parliament may by law provide that neither the Supreme Court nor any other court shall exercise jurisdiction in respect of any such dispute or complaint as is referred to in clause (1).

[30]The functioning of the Tribunals constituted under the Inter-State River Water Dispute Act, 1956 is a complex affair. Some of these Tribunals have been in existence for more than a decade or two. On the working of these tribunals and a personal account of working with them See Fali S. Nariman, "Inter-State Water Dispute: A Nightmare" in Ramaswamy R. Iyer (Ed.), *Water and the Laws in India* (Sage Publications: New Delhi, 2009) p. 32. Also see Radha D'Souza, "Nation vs. Peoples: Inter-State Water Dispute in India's Supreme Court" in the above book edited by Ramaswamy R. Iyer on p. 58.

[31]The Narmada Water Disputes Tribunal (October 1969) dealt with the sharing of Narmada Waters and Narmada River Valley Development (between Gujarat, Madhya Pradesh, Maharashtra and Rajasthan. The award was given on 7 December 1979. http://wrmin.nic.in/forms/list.aspx?lid=384&Id=4, Accessed on December 20, 2015; The Godavari Water Disputes Tribunal, constituted in April 1969, gave its award in July 1980. https://wrd.maharashtra.gov.in/portal-cms/cmspreview?v=1.1&l=en&p=/default/mwrd/homepage/new/godavari.html, Accessed on 20 December 2015; The Cauvery Water Disputes Tribunal finalised its award in 2007 and it is yet to be notified, http://wrmin.nic.in/forms/list.aspx?lid=378&Id=4, Accessed on December 20, 2015; The Krishna Water Disputes Tribunal award has been notified in 2010, http://wrmin.nic.in/writereaddata/Inter-StateWaterDisputes/KWDTReport9718468760.pdf, Accessed on December 20, 2015.

The Report of the Krishna Water Disputes Tribunal refers to the 1997 UN Convention. It also notes that this Convention has been made one of annexures in the affidavits filed by both Karnataka and Andhra Pradesh.[32] Articles 5 and 6 of the 1997 UN Convention have also been quoted in the text of the report. Besides this, several references have been made to the various cases decided by the US Supreme Court.[33]

2.4 Concepts of 'Cooperation' and 'Good Faith'

The Krishna Waters Disputes Tribunal while examining the factors relevant to settlement of disputes, *inter alia,* refers to what could be termed as a psychological element of "the feeling of cooperation and good faith amongst the concerned riparian States".[34] Apart from that, the Tribunal had further pointed out that "one of the clauses of the preamble of the Convention is about "affirming the importance of international cooperation and good neighbourliness in this field."[35] It is important to note that the Tribunal specifically referred to Articles 5 and 6 of the 1997 UN Convention, which dealt with equitable and reasonable utilization and participation.[36] The Tribunal, while concluding this argument essentially uses language that has reference to international legal principles such as "good faith", "mutual regard", and "mutual understanding". This could be regarded as an attempt by the Tribunal to juxtapose or bring in the principles of international law on river water sharing into an essentially domestic legal issue.[37]

2.4.1 Historical Context of Prescriptive Rights

The Cauvery Water Dispute Tribunal had to deal with several complex issues while apportioning waters between Karnataka, Tamil Nadu, Kerala, and Pondicherry. The

[32]See the Report of the Krishna Water Disputes Tribunal, p. 160.

[33]Ibid, pp. 167, 177, and 191.

[34]Ibid, p. 160.

[35]Ibid.

[36]Ibid, p. 167. The Tribunal takes note of the reference made by the State of Maharashtra to Article 5 and 6 of the United Nations Convention on Non-Navigational Uses of International Watercourses, 1997.

[37]Ibid.

dispute had a history of over a century.[38] The validity and legality of various arrangements and agreements concluded between then existing political regimes in the colonial context were also in question. There were issues relating to "prescriptive rights".[39] However, the Tribunal in its report begins with a general outline of what it intended to do while deciding on the issue of apportionment. The Tribunal noted that the "principles of apportionment of waters of inter-State or international rivers like principles of natural justice, have been evolved and developed by different courts from time to time in the course of more than a century while adjudicating the disputes between different States or Nations".[40]

The Cauvery Tribunal furthermore elaborates on the genesis of the water sharing issue and also the evolution of inter-state water disputes. It is worthwhile to quote some of these aspects as it seems to view the issue from a larger South Asian context:

> It is well known that most of the ancient cities and civilizations grew up on the banks of such rivers because of the fertile land and easy communication. But during the middle of 19th century because of the industrial revolution and allied development which brought prosperity to man-kind also gave birth to conflict and dispute in respect of sharing of waters

---

[38]Ibid. The Tribunal referring to the historical aspects, points out: "From records it shall appear that dispute about sharing of the water of river Cauvery is more than one and a half century old, details whereof have already been mentioned in earlier volumes. Before the Cauvery Fact Finding Committee, in the year 1972, claims had been made by different riparian States for 1,260.34 TMC (Ref: TNDC Vol. XV, page 110), whereas the aforesaid Committee as well as this Tribunal on consideration of different material adduced before this Tribunal have estimated the average yield at 50% dependability to be at 740 TMC".

[39]Ibid. On the issue of "prescriptive rights", the Tribunal noted the arguments of Tamil Nadu that being the lower riparian State it "…has a right of prior appropriation of the waters of the river Cauvery even in a proceeding relating to the apportionment of the waters of the said river." Tamil Nadu also referred to the Indus Commission of the year 1942 in which the Commission had pointed out that "priority of appropriation gives superiority of right; in general interest of the entire community inhabiting dry and arid territories; priority may usually have to be given to an earlier irrigation project over a later one." Tamil Nadu, quoting 1942 Indus Commission Report, noted that "the common law rule of riparian rights is completely destructive of equitable apportionment, for, under that rule, the upper owner can hardly take any share-far less his fair share-of the water of the river for purposes of irrigation. Therefore, that rule cannot be applied to an inter-State dispute even where it is recognized by both the States in their own internal disputes. The doctrine of appropriation, on the other hand, is consistent with equitable apportionment, provided that the prior appropriator is not allowed to exceed reasonable requirements. This condition is in fact part of the doctrine as enunciated by the Court in *Wyoming v. Colorado* [1922] (259 U.S. 419, 459) and again in *Arizona v. California* [1936] (298 U.S. 558, 566). Moreover, this doctrine is dictated by considerations of public interest; in arid territories where irrigation is a prime need, there would be no incentive for any individual or State to spend money upon an irrigation project, unless there was some assurance that it would not be ruined by subsequent diversion higher up the river. Where, therefore, both the States in an inter-State dispute recognize the doctrine of appropriation within their own borders, the most equitable course to apply that same doctrine to the determination of the dispute.

[40]*The Report Cauvery Water Dispute Tribunal with the Decision In the Matter of Water Disputes regarding the Inter-State River Cauvery and River Valley thereof.* See http://wrmin.nic.in/forms/list.aspx?lid=378&Id=4, Accessed on December 20, 2015.

of such inter-State and international rivers. If the history of such disputes in different parts of the world is examined, it will appear that sometimes the upper riparian States have been claiming an absolute right on the flow of water which used to pass through their territories. In other cases lower riparian States laid claim on the principle of right of easement saying that they have been enjoying the flow of that river for centuries and their economy is heavily dependent on such flows. As such there is no question of interrupting the flow of such river by the upper riparian State. This obviously led to disputes and disharmony in respect of sharing of waters by different States and nations and courts were faced with the situation how to strike a balance keeping the interests of all the riparian States. In some cases the matter was not so difficult while arriving at a reasonable and rational basis for sharing the water of an inter-State river because of the volume of the water available in the basin. The only question which was examined and answered was as to which State should get what proportion of water out of the total yield of the river. But the situation becomes grave and acute when the demands of the different States are much higher than the total available water in the basin in question.[41]

2.4.2 *Rule of Priority of Appropriation*

The "rule of priority of appropriation" was one of the major issues in the Cauvery Waters Dispute. The Tribunal noted the origin of this concept in the western States of the US and it was done through a statutory provision including incorporation of such rights into the respective Constitution of the States.[42] Referring to a study on the subject,[43] the Tribunal further noted that the appropriation system was an expedient means to encourage the development of the arid West, where much of the land is distant from streams and water was limited. The Tribunal also noted that the eight most arid States (Arizona, Colorado, Idaho, Montana, Nevada, New Mexico, Utah, and Wyoming) constitutionally or statutorily repudiated riparian rights very early and adopted prior appropriation as the sole method of acquiring rights to the use of water for all beneficial purposes.[44] The Tribunal, referring to various decisions of the US Supreme Court on this issue, concluded:

> The right of prior appropriation as existed in some of the Western States of USA cannot be equated with the past utilization of waters of the basin including existing utilization by one State or other in the inter-State basin. Past utilization or existing utilization has been recognized as a relevant factor in a proceeding for apportionment of waters of an inter-State or international river. It has its origin in point of a time, as to which of the State started the utilization first. The past utilization, which is also some time described as prior utilization, is a part of evolution and development of river basin linked with the history of the basin. As such the courts from time to time have taken that fact as a relevant factor while apportioning

[41]Ibid.

[42]Ibid. The Tribunal in its report noted that this was done in the US to help arid States to get enough water.

[43]Ibid. David H. Getches, *Water Law in a Nutshell* (3rd ed.), (University of Colorado School of Law: Boulder, Colorado, 1997).

[44]Ibid.

the water of an inter-State basin. But at the same time they have pointed out that some circumstances prevailing in the other riparian States may outweigh the prevailing practice and in that event such practice or use can be restricted or modified in a reasonable manner.[45]

2.4.3 *Rule of Equitable Apportionment*

Besides these issues, the Cauvery Tribunal Report dealt elaborately with the doctrine of "equitable apportionment" referring to various factors that should be taken into account. Some of these factors find elaboration both in Helsinki Rules and the 1997 UN Convention. The Tribunal, *inter alia,* noted the following:

> The doctrine of "Equitable Apportionment" cannot therefore be put in the narrow straight jacket of a fixed formula. In determining the just and reasonable share of the interested States, regard must be paid in the first instance to whatever agreements, judicial decisions, awards and customs are binding upon the parties. As to any supplies not controlled by these factors, the allocation may be made according to the relative economic and social needs of the interested States. The other matters to be considered include the volume of the stream, the water uses already being made by the States concerned, the respective areas of land yet to be watered, the physical and climatic characteristics of the States, the relative productivity of land in the States, the Statewise drainage, the population dependent on the water supply and the degree of their dependence, alternative means of satisfying the needs, the amount of water which each State contributes to the Inter-State stream, extent of evaporation in each State, and the avoidance of unnecessary waste in the utilization of the water by the concerned States.[46]

[45]Ibid. The Tribunal also referred to the views expressed in the Krishna Water Disputes Tribunal in Chapter XII (p. 98)—Existing use of a State is important evidence of its needs. Demands for potential uses are capable of indefinite expansion. Equitable apportionment can take into account only such requirements for prospective uses as are reasonable having regard to the available supply and the needs of the other States.

[46]Ibid. The Tribunal also refers to the views expressed by other tribunals such as Godavari Tribunal and Ravi Beas Waters Tribunal about the equitable apportionment of waters. From the Report of the Godavari Water Disputes Tribunal, in respect of the law of equitable apportionment, it has been observed at page 19 of Chapter IV: "In the absence of legislation, agreement, award or decree, the Tribunal has to decide the dispute in such a way as will recognize the equal rights of the contending States and at the same time establish justice between them. Equal right does not mean an equal division of the water. It means an equitable apportionment of the benefits of the river, each unit getting a fair share." In the Report of The Ravi Beas Waters Tribunal, in respect of claims of riparian States of an inter-State river it has been observed at page 94: "There is another reason which also militates against the view of the State owning proprietary rights in river waters. Even in ancient times flowing water was assimilated to the air and the sea. As a commodity it was common to all. A river was *res publica iure gentium,* open to navigation and fishing to all citizens. It was only feudal Lords who perhaps claimed absolute property rights over that part of the stream which crossed their territories. There is nothing in law for any one including the State to claim absolute proprietary rights in river waters. Running water has, therefore, rightly been called "a negative community" as it belongs to no one and is not susceptible to absolute ownership rights. The only right which a State can legitimately claim in river waters flowing within its territory is the right to make use thereof provided such use does not affect adversely the right which another State has to make use of the said waters."

The Tribunal further refers to and interprets the working and justifications for the three different views in respect of the claims by different riparian States regarding the sharing of the water of an inter-State river or a river passing from one nation to another. These are:

> the Harmon doctrine – according to this doctrine, the Tribunal notes, every State is sovereign and has the right to do whatever it likes with the waters within its territorial jurisdiction irrespective of the injury that it might cause to the neighbouring State by such appropriation and diversion;
>
> that a lower riparian State is entitled to water in its natural flow without any diminution or interference or alteration in its character. During the last century, as the Tribunal puts it, both views had been propounded – the first one by the upper riparian States and the second by the lower riparian States. If it is examined by an example, a State which is at the head of the river from which the river initially passes then such a State can utilize and divert the water from the said river making the lower riparian State starve, leading to the break-down of the economy of such lower riparian State. Similarly, if the second view is pushed to its logical end, then the upper riparian State, although it may be in dire need of the water of such inter-State river for agriculture and other uses, shall be a mute spectator of the water of such inter-State river flowing from its territory to the lower riparian State;
>
> the third view is based on the principle of "equitable apportionment", that is to say that every riparian State is entitled to a fair share of the water of an inter-State river according to its need. Such a river has been provided by nature for common benefit of the community as a whole through whose territories it flows, even though those territories may be divided by political frontiers.[47]

---

[47]The Tribunal refers to and discusses various decisions of the US Supreme Court spread over a century. In fact, it enters into a jurisprudential analysis of the US Supreme Court decisions. Some of these need mention in order to understand the working of the Tribunal basing its interpretative matrix on international and domestic decisions, especially within the US. These cases are: *Kansas* vs. *Colorado* (1906) and *Colorado* vs. *Kansas (*1943) where the US Supreme Court stated "The lower State is not entitled to have the stream flow as it would in nature regardless of need or use. If, then, the upper State is devoting the water to a beneficial use, the question to be decided, in the light of existing conditions in both States, is whether, and to what extent, her action injures the lower State and her citizens by depriving them of a like, or an equally valuable, beneficial use." In *New Jersey* vs. *State of New York* (283U.S.336) decided in 1931, the Court noted, "A river is more than an amenity, it is a treasure. It offers a necessity of life that must be rationed among those who have power over it. New York has the physical power to cut off all the water within its jurisdiction. But clearly the exercise of such a power to the destruction of the interest of lower States could not be tolerated. And on the other hand equally little could New Jersey be permitted to require New York to give up its power altogether in order that the river might come down to it undiminished. Both States have real and substantial interests in the River that must be reconciled as best they may be. The different traditions and practices in different parts of the country may lead to varying results but the effort always is to secure an equitable apportionment without quibbling over formulas…" In *State of Connecticut vs. Commonwealth of Massachusetts* {282 U.S.660} (1931) it was said: "For the decision of suits between States, federal, States and international law is considered and applied by this court as the exigencies of the particular case may require. The determination of the relative rights of contending States in respect of the use of streams flowing through them does not depend upon the same considerations and is not governed by the same rules of law that are applied in such States for the solution of similar questions of private right. The same question was considered in State of *Colorado vs. State of New Mexico*, by the US Supreme Court (1982). Justice Marshal who delivered the opinion on behalf of the Court said: "In addition, we

The above brief discussion of the analysis made by the various Indian Tribunals relating to various inter-State water sharing issues shows that even within the domestic context the applicable legal principles emanate from international law. Even within the context of international law, efforts are underway to codify norms relating to river water sharing. The 1997 UN Convention could be regarded as an attempt to codify the existing State practice at the inter-governmental level. It took more than two decades for the International Law Commission (ILC) to finalize these draft articles. This shows the difficulty in harmonizing and codifying the existing law on the subject. In the next part we will briefly examine evolution and application of international legal principles relating to river water sharing.

2.5 Evolution of River Water Sharing Norms Under International Law

The emerging international legal regime purports to lean on a more viable functional and equitable approach to river water sharing by referring to various 'uses'.[48] In other words, the sharing of the waters of a river would be directly linked to the kind of 'use' to which the river is being put by different stakeholders. Rivers have been historically used for navigational purposes and this use continues even

(Footnote 47 continued)

have held that in an equitable apportionment of inter-state waters it is proper to weigh the harms and benefits to competing States. In *Kansas v Colorado*, where we first announced the doctrine of equitable apportionment, we found that users in Kansas were injured by Colorado's upstream diversions from the Arkansas River. Yet we declined to grant any relief to Kansas on the ground that the great benefit to Colorado outweighed the detriment to Kansas. Similarly, in *Nebraska v. Wyoming*, we held that water rights in Wyoming and Nebraska, which under State law were senior, had to yield to the "countervailing equities" of an established economy in Colorado even though it was based on junior appropriations. We noted that the rule of priority should not be strictly applied where it "would work more hardship" on the junior user "than it would bestow benefits" on the senior user. The same principle is in balancing the benefits of a diversion for proposed uses against the possible harms to existing uses".

[48]United Nations Convention on the Law of the Non-Navigational Uses of International Watercourses, 1997 ("UN Convention" hereinafter) specifically refers to the term "equitable and reasonable utilization" in the context of use. Article 5 of this Convention, *interalia*, provides that "…In particular, an international watercourse shall be used and developed by watercourse States with a view to attaining optimal and sustainable utilization thereof and benefits there from, taking into account the interests of the watercourse States concerned, consistent with adequate protection of the watercourse". Article V of The Helsinki Rules on the Uses of the Waters of International Rivers ("Helsinki Rules" hereinafter) adopted by the International Law Association in 1966 refers *inter alia* to the term "reasonable and equitable share" and past utilization (including in particular existing utilization), comparative costs of alternative means of satisfying the economic and social needs and avoidance of unnecessary waste. See *Report of the Fifty-Second Conference*, International Law Association, Helsinki, 1966, p. 484.

today.[49] Rivers are also used for non-navigational uses such as irrigation, drinking, hydro-power generation, and towards other human requirements. While navigational use in the normal course does not diminish the quantity of water in a river, non-navigational uses would certainly diminish the quantity of water akin to natural factors such as evaporation and seepage. Navigational uses are generally regulated under territorial legal requirements and accordingly do not necessarily give rise to any major transboundary issues. The non-navigational uses of the international waters, on the other hand, are the most contentious ones as they affect the flow of the water from one State to another.[50]

It should, however, be noted that there is no inherent hierarchy in categorizing these 'uses'. For example, according to Article VI of the Helsinki Rules, "A use or category of uses is not entitled to any inherent preference over any other use or category of uses".[51] Article 10(1) of the 1997 UN Convention also provides that "In the absence of agreement or custom to the contrary, no use of an international watercourse enjoys inherent priority over other uses".[52]

---

[49]The UN Convention is specific to 'non-navigational' uses of the rivers. Chapter 4 of the Helsinki Rules addresses the 'navigational' uses of the rivers and lakes which extend to two or more States. It further provides for the "free navigation on the entire course of a river or lake" which, however, is subject to the exclusive jurisdiction of the riparian State.

[50]The Helsinki Rules were adopted by the International Law Association (ILA) in 1966. It, in fact, was an attempt to codify and existing international legal rules on the sharing of international rivers among States. Considering the status of ILA (as a body of lawyers and legal experts drawn from various parts of the globe and regarded essentially as a non-governmental body), the Helsinki Rules were regarded as having a less binding effect on States and had persuasive value under international law. The UN Convention, on the other hand, had been negotiated and finalized by the States *perse*. It was adopted in 1997 after a prolonged codification process undertaken by the International Law Commission (ILC) since 1974. The ILC took up this issue for consideration in 1974 pursuant to resolution 2669 (XXV) adopted by the UN General Assembly on 8 December 1970 on "Progressive Development and Codification of the Rules of International Law Relating to International Watercourses". The UN General Assembly had adopted resolution 1401 (XXIV) on 21 November 1959 pointing out that it was "desirable to initiate preliminary studies on the legal problems relating to the utilization and use of international rivers with a view to determining whether the subject is appropriate for codification". This codification process initiated by the UN finally resulted in the adoption of UN Convention on 21 May 1997. The UN Convention has come into force as on 17 August 2014. No South Asian country is a party to this Convention. See https://treaties.un.org/pages/ViewDetails.aspx?src=TREATY&mtdsg_no=XXVII-12&chapter=27&clang=_en. Accessed on 23 December 2015.

[51]*The International Law Association, Report of the Fifty-Second Session Conference*, n. 12, pp. 484–532. Also see a discussion on this issue in Surya P. Subedi, "Regulation of Shared Water Resources in International Law: The Challenges of Balancing Competing Demands" in Surya P. Subedi (Ed.) *International Watercourses Law for the 21st Century*, n. 2, pp. 7–18. Subedi points out "However, there are a number of areas around the world where the existing bilateral or multilateral agreements have not been able to provide satisfactory solutions to the problems faced by the watercourse States of the region".

[52]Article 10 of the UN Convention on the Law of the Non-Navigational Uses of International Watercourses, 1997 seeks to define the "Relationship between different kinds of uses" and it further provides in clause (2) that "In the event of a conflict between uses of an international watercourses, it shall be resolved with reference to articles 5 to 7, with special regard being given

Considering the complexities involved in defining the term 'use', States usually prefer to resolve such issues bilaterally. Accordingly, States have been concluding bilateral or regional river water sharing arrangements defining the sharing formula usually based on the extent of the 'use' of the river water basin.[53] It should be noted that nearly half of the global population lives in river basins cutting across, in several cases, national boundaries.[54] It should be further noted that States, while disregarding the "river-basin approach,"[55] have attempted to look at the issue of sharing of waters of the international rivers from the perspective of "absolute territorial sovereignty" principle.[56] It is sometimes argued that a State located on the upstream termed as "upper riparian" would, on account of its geographical location, have priority right of access to river waters when compared to a State located downstream termed as "lower riparian." This was the legal position taken by the United States in 1895 and came to be known as the Harmon Doctrine.

2.5.1 *The Harmon Doctrine*

The Harmon doctrine supported the legal position of an upper riparian State to have access and use the waters of international rivers in absolute terms. This doctrine was propounded by the Attorney General Judson Harmon of the United States in 1895 in a legal opinion given to the US State Department. The case concerned a dispute between the US and Mexico with regard to sharing the waters of the Rio Grande River for irrigational purposes. Basing his legal opinion on earlier US Supreme Court decisions, Attorney General Harmon stated that the US had sovereign rights over the resources located within its boundaries. As the water of the Rio Grande

(Footnote 52 continued)

to the requirements of vital human needs". For the text of the 1997 UN Convention see https://treaties.un.org/doc/Publication/MTDSG/Volume%20II/Chapter%20XXVII/XXVII-12.en.pdf, Accessed on 23 December 2015.

[53]Article 3 of the UN Convention refers to the Water Course Agreements and explicitly recognizes the rights of the States to conclude such agreements. However, it requires States to "consider harmonizing such agreements with the basic principles of the present Convention".

[54]Stephen C. McCaffrey, *The Law of International Watercourses: Non-Navigational Uses* (Oxford University Press: New York: 2001) p. 8.

[55]Ibid, The river basin approach seeks to strike a balance between the existing natural functions of the river system and societal expectations for livelihood, industry, recreation, nature management and agriculture. For a brief account on this aspect generally see http://documents.Worldbank.org/curated/en/965371468340137430/pdf/411500Intro0to1mgmt0NOTE1101PUBLIC1.pdf, Accessed on 23 December 2015.

[56]Ibid, p. 112. McCaffrey lists four principal theories relating to the non-navigational uses of the international watercourses: absolute territorial sovereignty, absolute territorial integrity, limited territorial sovereignty, and community of interests. Absolute territorial sovereignty principle emanated from the Harmon Doctrine wherein a State had the complete freedom to decide the use of the portion of its international river.

flowed within US, he argued, the US had absolute rights to use it the way it wanted.[57]

The Harmon doctrine, as noted by several international legal scholars, did not survive, although some States, especially upper riparians, continue to argue and exercise their rights even today in line with Harmon Doctrine.[58] The US, in fact, repudiated the Harmon Doctrine when it dismissed Canada's reliance on Article II of the 1909 treaty with regard to sharing of waters of the Columbia River. The US argued that the Harmon Doctrine was not part of international law. Both Mexico and US eventually entered into a dialogue process to work out the sharing of the waters of the Rio Grande River and concluded a joint mechanism called the 1906 Convention concerning the Equitable Distribution of the Waters of the Rio Grande for Irrigation Purposes. The idea of concluding a Joint Commission (comprising experts from both States) emerged from this 1906 agreement between US and Mexico.

2.5.2 Joint Mechanisms and Joint Commissions

The domestic practice of United States since the latter part of the nineteenth century has provided rich jurisprudence towards the evolution of laws relating to river water

[57]The Harmon Doctrine, *inter alia*, provides that "a State has complete freedom of action with regard to the portion of an international watercourse that is situated within its territory, irrespective of any harmful consequences that may ensue for other riparian States." The length of the Rio Grande River is about 3,000 km and half of this length forms the boundary between the United States and Mexico. It is also important to note that preceding this 1895 legal opinion by the U. S. Attorney General, Harmon both the contending States entered into a prolonged negotiations about the flow of water for nearly five years. Even the US Congress had agreed in 1890 through a resolution that upstream diversions from Rio Grande were depriving those on the Mexican side of water. Both the houses of Congress had requested the US President to enter into negotiations with Mexico with a view to resolving the Rio Grande water problems. Mexico had also referred to "the principles of international law" and argued they "would form a sufficient basis for the rights of the Mexican inhabitants on the bank of the Rio Grande". Mexico, further referred to, Article VII of the 1848 Treaty of Guadalupe Hidalgo according to which both the countries had an obligation not to construct any work without the consent of the other that may impede or interrupt, in whole or in part, the flow of the water in the river. Harmon's legal opinion sought to nullify these arguments by stating that treaty provisions do not directly address private citizens and that it only applies to governments. These issues also brought into focus the domestic implementation of international law and treaties within the U.S. Later, the U.S. Government in a case before the Supreme Court (*United States v. Rio Grande Dam & Irrigation Co.*) concerning these issues on the Rio Grande River did not concur with the legal opinion given by the Attorney General Harmon. For a detailed account on this issue see McCaffrey, *supra* n. 55, pp. 76–111.

[58]Ibid. According to McCaffrey "A historical survey of the views of commentators shows that while there was some support for the theory of absolute territorial sovereignty in the nineteenth century and even in earlier in decades of twentieth century, it declined sharply as the significance of non-navigational uses increased".

sharing.[59] As mentioned above, the idea of a Joint Commission emerged from the 1906 US-Mexico Agreement. The joint mechanisms and commissions provide flexibility to States to negotiate and allow for constant monitoring of the situation. In sharing the waters of the Colorado River with Mexico, the idea of a Joint Commission was formalized in 1944 in a Treaty between the US and Mexico relating to the Utilization of the Waters of the Colorado and Tijuana Rivers, and of the Rio Grande (Rio Bravo) from Fort Quitman, Texas to the Gulf of Mexico. The Colorado River has been described as the most legislated, most debated, and most litigated river in the entire world.[60] The Columbia River flowing as a boundary river between the United States and Canada was the subject of a treaty, namely the 1909 Treaty Relating to Boundary Waters and Questions Arising between the US and Canada. The Treaty established the International Joint Commission of the United States and Canada through which several disputes related to water sharing have been resolved. But for nearly four decades both the US and Canada had bitter diplomatic conflicts with regard to sharing of the waters of the Columbia River. This dispute between Canada and the US was finally resolved with the conclusion of the 1961 Treaty relating to Cooperative Development of the Water Resources of the Columbia River Basin.[61]

---

[59]The Indian team that was negotiating the Indus Treaty with Pakistan in the decade of 1950s visited various parts of the United States several times to understand the legal and technical aspects of the river sharing arrangements. See Gulati, n. 7, p. 9. The US practice was studied and closely examined before incorporating water sharing arrangements that now exist in Article 262 in the Indian Constitution. However, this procedure under the Indian Constitution has been critiqued, specifically in excluding the jurisdiction of the Supreme Court. Fali Nariman points out that “In USA, such disputes remain with the country’s Supreme Court, a procedure having been devised for appointing a Special Master to record evidence and give his findings on various uses of fact and law, and make his report to the Court; each of the contesting States then filing objections or responses to the findings in the report, and the Supreme Court of the United States ultimately pronouncing its final verdict”. He further notes that “I believe it was an error for us to have departed from the American pattern (the US has a written Constitution like ours) to resolve inter-State river water disputes. When conceived way in the year 1956, it could perhaps have been justified as an innovative experiment-but the experiment has been a failure” (Fali Nariman, n. 31, p. 38–39). Article 33 of the UN Convention also reflects this idea. It, *interalia*, provides for the settlement of disputes by States through peaceful means. If they fail, “they may jointly seek the good offices of, or request mediation or conciliation by, a third party, or make use, as appropriate, of any joint watercourse institutions that may have been established by them or agree to submit the dispute to arbitration or to the International Court of Justice”. Article 33 also provides for Fact-finding Commissions comprising members from both parties. Article 8 of The Indus Water Treaty, 1960, for example, provides for the Permanent Indus Commission that seeks to resolve all bilateral issues through ongoing negotiation.

[60]See Mark Reisner, *Cadillac Desert: The American West and Its Disappearing Water* (Penguin 1993).

[61]This Treaty provides for comprehensive cooperation between the US and Canada in planning and sharing the river. Canada agreed to construct several large storage dams to benefit the US to enhance its power generation capabilities and as protection against floods. The US, on its part, agreed to provide Canada with one-half the additional power and to compensate Canada for flood-control benefits. According to McCaffrey “…while Canada had initially taken a Harmon-Doctrine-type position, based upon Article II of the 1909 treaty, in the end the two states

The US experience in river water sharing has shown that the appropriate way to resolve the river conflicts would be to put in place a joint mechanism comprising all the parties to the dispute. This kind of mechanism could monitor the situation from time to time and suggest necessary adjustments. International law, as it evolved during this period, preferred negotiated settlement of all river sharing conflicts. This US practice with regard to non-navigational uses of international rivers has influenced the content and formation of various bilateral and regional treaties.[62]

The 1960 Indus Water Treaty between India and Pakistan, which was finalized after almost decade-long negotiations, borrowed many legal formulations from the US's internal state practices.[63] For example, the Indus Treaty provides for the Indus Commission comprising two Commissioners who would meet periodically and report back to their Governments.[64] They are supposed to discuss and implement the day-to-day requirements relating to such things as exchange of information, periodic data on release of water, and other procedural requirements under the Treaty. For all other substantive issues Indus Commissioners have to consult their Governments. The 1996 Treaty on Sharing of the Ganga/Ganges Waters at Farakka between India and Bangladesh also provides for a Joint Committee consisting of an equal number of members nominated by the two Governments.[65] The 1996 Mahakali Treaty between India and Nepal also provides for the Mahakali River

---

(Footnote 61 continued)

arrived at an equitable apportionment of the benefits of the Columbia River" (McCaffrey, n. 55, p. 296).

[62]Besides bringing in various legal and technical principles from the US Supreme Court decisions, the idea of creation of Tennessee Valley Authority (TVA) in 1933 by the US Congress was adopted by India as well. The US Congress created TVA as a federal authority in the times of Great Depression to address wide range of environmental, economic and technological issues, including the delivery of low cost electricity and the management of the natural resources. TVA covers an area of about 80000 sq. miles and its benefits spreading over several of the states of the US; see generally on the creation and current functioning of TVA see http://www.tva.gov/. The World Bank which facilitated the negotiation of the Indus Water Treaty had proposed a similar such body on the lines of TVA between India and Pakistan. Later, India took up this TVA idea and created Damodar Valley Corporation (DVC) by an act of Central Legislature, namely Damodar Valley Corporation, 1948 (Act No. XV of 1948). It should be noted that the creation of DVC was preceded by a study prepared by W. L. Voorduin, a senior engineer of TVA in 1944. This study entitled "Preliminary Memorandum on the United Development of the Damodar River" provided the base for the creation of DVC consisting of eastern and northern Indian states; see http://www.dvcindia.org.

[63]Gulati, n. 3, p. 15.

[64]Article VIII of The Indus Water Treaty according to which Permanent Indus Commissioners for all matters arising out of the Treaty would serve as the regular channels of communication between the two Governments (unless Governments decide take up any particular question directly with each other).

[65]Articles 4–7 provide for the creation and working of a Joint Committee. This Committee will be responsible for implementing the arrangements contained in the Farakka Treaty and examining any difficulty arising out of the implementation of the above arrangements and of the operations of Farakka Barrage.

Commission comprising an equal number of representatives from both States.[66] As stated earlier, the Indian Constitution does not allow river water sharing issues to be adjudicated directly in the courts. It requires the Government to constitute a tribunal for this purpose. One could safely assert that even this provision of the Indian Constitution with regard to river water sharing was influenced by the existing State practices, specifically of the US.

2.5.3 *Efforts Towards Codification*

The process of codification of the state practices relating to water sharing evolved over a very long period of time.[67] Till the adoption of the 1997 United Nations Convention on the Non-Navigational Uses of the International Watercourses, there were no inter-governmental efforts to bring into existence a multilateral treaty on the subject of river water sharing. There were, however, two major non-governmental efforts to codify or restate the existing law on international rivers. This codification by a non-governmental body would only reflect the existing nature of law without any binding commitments on the part of States. It has also been argued that the resolutions or declarations passed by these bodies did not entirely reflect existing state practices.[68]

2.5.4 *The Institute of International Law (L'Institut de Droit International)*

Various resolutions and declarations passed by the Institute of International Law (*L'Institut de Droit International*) on Law of International Rivers since 1911 have

[66]Article 9 of the Treaty provides for the creation of a Mahakali River Commission, which will be "guided by the principles of equality, mutual benefit and no harm to either party".

[67]The Institut de Droit International began its formal work on water sharing with its Madrid Resolution of 1911. Similarly, the Helsinki Rules were adopted in 1966. The debate within the United Nations General Assembly to codify the existing State Practice began in 1970. Later, the International Law Commission took up the work in 1974 and submitted its final draft to the General Assembly in 1994. The same was adopted by the General Assembly after prolonged discussion within the Sixth Committee during May 1997.

[68]The General Assembly of the United Nations debated this issue in its 25th Session in December 1970. Turkey and other countries, while noting the work done by the International Law Association and others, argued for taking up the codification of existing State practice on sharing of international watercourses. Accordingly, the UN General Assembly noted this issue in its resolution. Res. 2669 (XXV), *ProgressiveDevelopmentandCodification of the Rules of International Law relating to International Watercourses*, 8 December 1970. Available at https://documents-dds-ny.un.org/doc/RESOLUTION/GEN/NR0/349/34/IMG/NR034934.pdf?OpenElement. Accessed on December 20, 2015.

an immense persuasive value under international law.[69] The *Institut de Droit International*, established in 1873, has been regarded as one of the most influential international institutions. Its members are eminent international lawyers who meet once or twice a year in different parts of the world and codify their views through resolutions and declarations. First such effort relating to river water sharing was made in 1911 by *Droit International* through its Madrid Resolution on International Regulations regarding the Use of International Watercourses. According to one scholar, the Madrid Resolution was remarkable not least for the principles it recognized in the early years of the twentieth century, when the law of the non-navigational uses of international watercourses was still in its formative stages.[70] This resolution, for the first time, noted that the international law had dealt with navigation but not "the use of water for the purposes of industry, agriculture etc." It provided for the first time that neither State on a contiguous watercourse may, without the other's consent, allow changes to be made to the watercourse that are detrimental to the opposite bank. It also provided that neither State may utilize the water in such a way as to interfere seriously with other State's utilization thereof.

It should be noted that this restatement of the law by *Droit International* provided one of the basic concepts of international law i.e., the obligation not to cause transboundary harm. This principle has been applied in two other well known cases, namely *The Trail Smelter Arbitration*[71] between US and Canada (1939) and *The Lake Lanoux Arbitration*[72] (1957) between France and Spain.

The 1961 Salzburg Resolution by *Droit International* for the first time introduced the concept of river basin as a criterion for river water sharing. This

---

[69]For the work of The *Institut de Droit International* and the three resolutions passed by it on international rivers, see www.idi-iil.org/. Accessed on December 20, 2015.

[70]McCaffrey, *supra* n. 55, pp. 318–320; Further, the work of the *Droit International* is summed up McCaffrey as "…a trend towards applying legal rules to the entire hydrographic basin rather than merely to the surface water channel; one in favour of increased use of procedural rules, possibly culminating in the establishment of joint management mechanisms; and a trend towards the idea that it is in the interest of all riparians that shared water resources be utilized in an equitable and reasonable manner".

[71]This Arbitration was concluded in 1939. This arbitration was about the transboundary pollution —with regard to a smelter that was located within Canada in a place called Trailer; but this smelter caused damage to crops within the territory of the United States. The Arbitration, *inter alia*, held that "…no State has the right to use or permit the use of its territory in such a manner as to cause injury by fumes in or to the territory of another or the properties or persons therein" (McCaffrey, n. 55, p. 206).

[72]Ibid. This arbitration was decided in 1957 between France and Spain. This was about a proposed hydro-electric power project by France and the same was opposed by Spain as it involved diversion of Lake Lanoux waters. The Arbitral Tribunal, as McCaffrey points out, *inter alia* referred to several principles of general international law such as (a) prohibiting the upper riparian State from altering the waters of a river in circumstances calculated to do serious injury to the lower riparian State; (b) upper riparian State in good faith to take into account the concerns of lower riparian; (c) obligation to take adverse interests into account in the course of negotiations and the obligation to give a reasonable place to such interests in the solution adopted.

resolution referred to "the utilization of waters which form part of a watercourse or hydrographic basin which extends over the territory of two or more States." It also provided for the first time what has been termed as the concept of "equitable utilization" when it stated that any dispute as to the extent of the respective States' rights "shall be settled on the basis of equity, taking into consideration the respective needs of the States, as well as any other circumstances relevant to any particular case". The Salzburg Resolution also provided for the first time for advance notice of new uses and negotiations in the event of objections to such uses.

2.5.5 Helsinki Rules

The Helsinki Rules on the Uses of the Waters of International Rivers was adopted in 1966 by the International Law Association (ILA).[73] Like *Droit International*, ILA is also a non-governmental body comprising lawyers from different parts of the world. It should be noted that the 1966 Helsinki Rules expanded the legal framework relating to river basin or drainage basin and also put in place guiding principles on equitable utilization. The Helsinki Rules defined an "international drainage basin" as a "geographical area extending over two or more States determined by the watershed limits of the system of waters, including surface and underground waters, flowing into a common terminus".[74] This definition, as noted by McCaffrey, is noteworthy not only for its broad approach, which is consistent with hydrological reality, but also for its specific mention of "underground waters".[75] He further and correctly points out that this increasingly important source of freshwater i.e., underground waters had largely escaped international legal regulation up to this point. This could be regarded as an important contribution by the Helsinki Rules.

However, the broad definition of "drainage" basin adopted by the ILA has been criticized as it placed heavy burden on upstream States.[76] States, specifically upper riparians, have not generally accepted the basin approach to river water sharing. The discussions within the Sixth Committee of the UN General Assembly have noted that the basin concept would put too much emphasis on the land areas within the watershed and that the physical land area of a basin might be governed by the

[73]International Law Association (ILA) also took more than a decade to codify these Rules. ILA had constituted in 1954 a Committee on the Uses of Waters of International Rivers to elaborate general rules applicable to international rivers drawing on the bilateral and regional practice and experience of States.

[74]Article II of The Helsinki Rules on the Uses of the Waters of International Rivers, International Law Association, *Report of the Fifty-Second Conference*, Helsinki, 1966, p. 484.

[75]McCaffrey, *supra* n. 55, p. 320.

[76]Malgosia Fitzmaurice and Gerhard Loibl, "Current State of Development in the Law of International Watercourses" in Surya P. Subedi (Ed.), *supra* n. 6, p. 35.

rules of international water resources law.[77] It should be noted that the Helsinki approach has been followed in several river water sharing treaties such as, for example, the 1972 Senegal River Basin Treaty, the 1987 Zambezi River Treaty, the 1978 Treaty on Amazonian Co-operation, the 1994 Convention on Co-operation for the Protection and Sustainable Use of the Danube River, and the 1995 Agreement on the Co-operation for the Sustainable Development of the Mekong River Basin.

One of the most important aspects of The Helsinki Rules relates to its emphasis on the equitable utilization of the waters that are available within the international drainage basin. The Helsinki Rules also provided for each Basin State, within its territory, to have a reasonable and equitable share in the beneficial uses of the waters of an international drainage basin.[78] Further, it also provided for the non-exhaustive list of factors to be taken into account in determining what amounts to a "reasonable and equitable share" in a specific case.[79] The factors to be considered are: (a) geography of the basin; (b) hydrology of the basin; (c) climate; (d) past utilization of the waters; (e) economic and social needs of each basin State; (f) population dependent upon the waters of the basin; (g) the comparative costs of alternative means of satisfying the economic and social needs of the each basin State; (h) availability of other resources; (i) avoidance of unnecessary waste in the utilization of the waters of the basin; (j) practicability of compensation to one or more of the co-basin States as a means of adjusting conflicts among uses; and (k) avoidance of substantial injury to a co-basin State. The causing of harm to another State is not prohibited, but is dealt with as a factor to be taken into account in determining whether a use is equitable. The Helsinki Rules also articulated guidelines for addressing issues related to pollution and compensation[80] and outlined procedures for dispute settlement.[81]

India had some reservations about the Rules although these have not been articulated in official terms. Reference has been made to Nagendra Singh's lengthy paper rejecting the notion of equitable utilization as mentioned in the Helsinki Rules on the grounds that "the ideas regarding the concept of equitable sharing are not clear" and that "States must be free to develop their uses in accordance with their needs".[82] To sum up, as noted by one scholar, "Helsinki Rules obviously constitute a monumental work. They have had a major impact upon the

[77]Ibid; also see *Yearbook of the International Law Commission* II (1983), Part One pp. 167–168.

[78]Article IV of the Helsinki Rules.

[79]Article V of the Helsinki Rules provide for all the relevant factors that could be helpful in determining the reasonable and equitable share.

[80]Chapter 3—Articles IX to XI.

[81]Chapter 6—Articles XXVI to XXXVII.

[82]Quoted by B. S. Chimni, "A Tale of Two Treaties: The Ganga and Mahakali Agreements", in Surya P. Subedi (Ed.), n. 2, p. 93: It has been further pointed out that despite these kinds of ideas, the Helsinki Rules had been applied by tribunals in India in adjudicating interstate water disputes.

development of the law of international watercourses, and reflect many principles and trends that later found expression in the UN Convention".[83]

2.5.6 *The 1997 United Nations Convention*

Several States, specifically upper riparian States, were opposed to the broad approach taken by the Helsinki Rules in defining an "international drainage basin". They argued that this definition failed to reflect the actual hydrographic reality of the time. Although the Helsinki Rules codified a large and heterogeneous body of international legal principles relating to international rivers, these were not binding on States. As noted above, the ILA under whose auspice these rules were adopted was a non-governmental body. Therefore, a proposal was made by Turkey and others within the Sixth Committee of the UN General Assembly to study this topic of "law of international rivers" for codification. In 1972, the UN International Law Commission formally adopted this topic for study. It took another 25 years for the UN to finalize and adopt a Convention on Non-Navigational Uses of International Watercourses.

This Convention was put to a vote in the UN General Assembly on 21 May 1997. While 103 States voted for it, 3 States opposed it. 27 States, including India and Pakistan, abstained. Bangladesh and Nepal voted for it.[84] Although it entered into force on 17 August 2014, none of the South Asian countries have joined the Convention as of date. China was among those three countries along with Turkey and Burundi to oppose this Convention. India abstained saying that the Convention failed to take into account sufficiently the existing regimes on water sharing (Article 3). It was also opposed to fact-finding missions proposed in Article 33 of the Convention. It consistently opposed giving access to nationals of other co-riparian States to pursue private law remedies within India.[85] China voted

---

[83]McCaffrey, n. 5, p. 321; It has been noted by McCaffrey that the Judge E.J. Manner from Finland who had chaired the ILA Committee that prepared the Helsinki Rules and also as a official Finnish delegate to the UN General Assembly had proposed at the UN that the International Law Commission take up the study of the law of the non-navigational uses of international watercourses and that it consider using the Helsinki Rules as a model for its work. However, as McCaffrey points out that political considerations prevented this and this proposal was not included. Some States within the UN General Assembly were opposed to using the Helsinki Rules as a basic model for initiating the codification process.

[84]*Convention on the Law of the Non-navigational Uses of International Watercourses*, A/RES/51/229, United Nations General Assembly, 8 July 1997.

[85]Firstly, India argued that Article 3 (concerning rights and obligations under the existing bilateral watercourse agreements) failed to adequately reflect the principle of freedom, autonomy and the right of States to conclude international agreements on the international watercourses without being fettered by the present Framework Convention. Secondly, Article 5 (about equitable and reasonable utilization and participation) had not been drafted in clear and unambiguous terms stating the right of the State to utilize an international watercourse for non-navigational purposes in

against as it felt that the Convention was imbalanced as it primarily favoured lower riparian states. It should be noted that the negative votes of China and Turkey were probably attributable to their positions as upstream States in an ongoing controversies such as China's plans to construct additional dams on its rivers, specifically upper Mekong. Similar sentiments were expressed by Turkey with specific reference to groundwater sharing.

Some of the salient features of the 1997 UN Convention could be highlighted, *albeit* briefly. These are:

> It moved away from the "international drainage basin" concept to a "watercourse system". It defines "watercourse" as "a system of surface waters and ground waters constituting by virtue of their physical relationship a unitary whole and normally flowing into a common terminus".[86]
>
> It recognized the validity, rights, and obligations undertaken by States in other watercourse agreements. However, it required that States, where necessary, consider harmonizing existing bilateral agreements with the basic principles of the UN Convention.[87]
>
> The obligation not to cause significant harm is an important principle under the Convention.[88] It also places an obligation on the Watercourse State to take all appropriate measures (with due regard to the provisions of reasonable and equitable utilization) in consultation with an affected State to eliminate or mitigate such harm and where appropriate to discuss the question of compensation.[89] There is also a general obligation to cooperate on the basis of sovereign equality, territorial integrity, mutual benefit, and good faith. There is also a provision to consider the establishment of joint mechanisms or commissions.
>
> The regular exchange of data and information on the condition of the watercourse, in particular that of a hydrological, meteorological, hydrogeological, and ecological nature and related to the water quality as well as related forecasts is also part of the Convention.[90] This should be done on a regular basis by the Watercourse State. If data/information is not readily available, the concerned Watercourse State should employ its best efforts to comply

(Footnote 85 continued)

an equitable and reasonable manner. Article 33 dealing with settlement of disputes contained in it an element of compulsion in so far as it envisaged the creation of a fact-finding commission. India further noted that any procedure for peaceful settlement of disputes should leave the parties to the dispute to choose freely and by mutual consent a procedure acceptable to them. It was opposed to the imposition of any mandatory third party dispute settlement procedure on a State without its consent. India felt that Article 5 in the present form is vague and difficult to implement. Thirdly, India had objections to Article 32 as it presupposed political and economic integration among States of the region. According to India as all watercourse regions are not so integrated, this provision will be difficult to implement in certain regions. Hence, it argued for its deletion (Indian statement while abstaining from voting at the UN General Assembly on 21 May 1997—on file with the author).

[86]Article 2 of the 1997 UN Convention on Non-Navigational Uses of International Watercourses defines various terms in the Convention.

[87]Article 3, 1997 UN Convention on Non-Navigational Uses of International Watercourses.

[88]Article 7, 1997 UN Convention on Non-Navigational Uses of International Watercourses.

[89]Article 8, 1997 UN Convention on Non-Navigational Uses of International Watercourses.

[90]Article 9, 1997 UN Convention on Non-Navigational Uses of International Watercourses.

with the request. It may also ask for payment of reasonable costs of collecting and processing such data and information.[91]

The possible effect of any planned measure on the watercourse also should be taken into account through the exchange of information and consultation with each other. If found necessary affected Watercourse States should engage in negotiations about the possible effects of planned measures on the condition of the international watercourse.[92] Further, notification concerning planned measures with possible adverse effects accompanied by available technical data and information, including the results of any environmental impact assessment should also be taken into account by the watercourse States. These measures, as per the Convention, will assist in evaluating the possible effects of the planned measures.[93] If the Watercourse States fail to notify each other about planned measures, and if one of them has reasonable grounds to believe that such measures are being carried out and that they are having an adverse effect, both the States have an obligation to enter into immediate consultations and negotiations exchanging all the available documentation.[94]

The protection and preservation of ecosystems of international watercourse forms an important element of the Convention.[95] The "pollution of an international watercourse" has been regarded as having taken place when there is "a detrimental alteration in the composition or quality of the waters of an international watercourse which results directly or indirectly from human conduct".[96] Watercourse States can also take measures necessary to prevent the introduction of species, alien or new, into an international watercourse, which may have effects detrimental to the ecosystem of the watercourse resulting in significant harm to other Watercourse States. The protection and preservation of the marine environment, in particular estuaries, are also provided for in the Convention. Such protection of marine environments should generally be done taking into account the already existing international rules and standards, such as the United Nations Law of the Sea Convention and other related maritime conventions relating the preservation and management of environment.[97]

The UN Convention does not oblige a Watercourse State to provide data or information vital to its national defence or security.[98] However, a State is free to provide as much information as it can under the circumstances. The Convention also envisages allowing a

---

[91]Ibid.

[92]Article 11, 1997 UN Convention on Non-Navigational Uses of International Watercourses.

[93]Strict time frames have been proposed for the reply to notifications. Article 13 provides for a six-month limit to the notified State to study and evaluate the possible effects of the planned measures. The notifying State also has an obligation to not to implement or permit the implementation of the planned measures without the consent of the notified States (Article 14).

[94]Article 18, 1997 UN Convention on Non-Navigational Uses of International Watercourses.

[95]Article 20, 1997 UN Convention on Non-Navigational Uses of International Watercourses.

[96]Article 21, 1997 UN Convention on Non-Navigational Uses of International Watercourses; Watercourse States can individually, where necessary, and jointly take action to reduce harm to their environment, including harm to human health or safety. Watercourse States, if need be, can harmonize their policies in this regard.

[97]United Nations Law of the Sea Convention, 1982, for example, has provisions relating to environment and pollution. Articles 192–237 under Part XII (Protection and Preservation of Marine Environment).

[98]Article 31, 1997 UN Convention on Non-Navigational Uses of International Watercourses. It is not clear as to who will decide this criterion of "national security". One argument could be—it will have to be decided by the Watercourse State itself, taking into account its own circumstances. This could be a very broad argument and States could bring any of the water-related issues under the

Watercourse State that has been seriously affected or that has suffered significant transboundary harm due to another Watercourse State to take recourse to the legal system or obtain access to judicial or procedures of that State to claim compensation or other relief.[99] Even for the settlement of disputes, all options such as negotiations, good offices, arbitration and other third party settlement mechanisms are available. If the dispute is not solved even after all these means and mechanisms, the Convention provides for the creation of an "impartial fact-finding" Commission.[100]

The verdict in the case concerning the *Gabcikovo-Nagymaros* decided at the World Court just before the adoption of the 1997 UN Convention is also crucial as it reiterated some of the normative aspects of the 1997 UN Convention.[101] The case concerned a large project on the Danube River pursuant to a Treaty between Hungary and Czechoslovakia (now Slovakia). As per the 1977 Treaty, the project consisted of a series of dams and barrages on a stretch of approximately 200 kilometres of the Danube between Slovakia and Hungary. Both Hungary and Slovakia brought this case to the World Court in July 1993. The Court delivered its judgment in September 1997. It has been pointed out that the Court's invocation of the Convention constituted a strong endorsement of the Convention as an authoritative instrument in the field and seems likely to lead States to refer to it in support of their position concerning internationally shared water resources.

2.6 Conclusion

The sharing of water resources is one of the most contentious issues among stakeholders. The present study is an attempt to examine these aspects in two contexts. The first context relates to the domestic legal systems and explores how these systems have grappled with this issue of amicably arriving at sharing river water resources. The ways in which the constitutional regimes of the South Asian countries deal with river water sharing challenges remain the primary focus of the first part of this study. This analysis underlines many similarities among these

(Footnote 98 continued)

broad rubric of vital national security. There appears to be no third party determination of this issue.

[99]Article 32, 1997 UN Convention on Non-Navigational Uses of International Watercourses. India had reservations on this provision. The phrase `other relief' could include injunctions or other similar legal means or measure to obtain a relief within the jurisdiction of another country.

[100]Article 33, 1997 UN Convention on Non-Navigational Uses of International Watercourses. Those Watercourse States which are Parties to the dispute have an obligation to provide access to the Fact-Finding Commission to the respective territories and to inspect any facilities, plant, equipment, construction or natural feature relevant for the purpose of its enquiry. India had reservations on this provision as well. It should be noted that there is a separate Appendix to the Convention dealing with Arbitration procedures.

[101]For the full details and text of the case see http://www.icj-cij.org/files/case-related/92/092-19970925-JUD-01-00-EN.pdf, Accessed on December 23, 2015.

countries in terms of their legal and policy approaches towards sharing water resources. These countries share a historical and cultural heritage, which is in some ways reflected in their factual realities, as well as their legal and policy structures as well.

All the South Asian countries have incorporated a reference to the need for the protection and preservation of environment and natural resources in their respective Constitutions. This broader Constitutional reference would include the rivers and watercourses as well within its interpretative orbit. The majority of the States around the globe, including all the South Asian States, seek to avoid any formal and binding adjudicatory process to settle the river water sharing differences so as to balance differing uses and interests. India and Pakistan clearly seek to exclude in their respective Constitutions itself the jurisdiction of their highest Courts in intervening in the substantive aspects of the water sharing issues. On the contrary, upon failing to resolve the water sharing issues through negotiations, states and the federal units (or the provinces) have taken recourse to the highest courts to trigger the dispute resolution mechanisms.

The second part of the study relates to the study and evolution of international legal principles. The international law on river water sharing has a history of over hundred years. Initially, the US practice and the decisions of its Supreme Court contributed to the development of several legal principles and criteria relating to equitable apportionment of river water. Besides the court decisions within the US, several river water sharing arrangements and joint mechanisms that had been developed among various States of the US have also contributed to the development of both law and policy towards this area. The earliest codification efforts in identifying and conceptualizing the existing state practices and other related practices among States was initiated by the *Institut de Droit International.* The work of the *Institut* had been regarded as forward-looking and provided a solid basis for future work. The work of the International Law Association (ILA) in a report submitted to its 52nd Session at Helsinki outlined the basic legal principles relating to the river water sharing. The Helsinki Rules essentially took a "basin approach" towards resolving the river water sharing arrangements.

The work of the *Institut* and the Helsinki Principles, though widely acknowledged and used by various stakeholders, has been regarded as the work of the non-governmental bodies failing in the process to sufficiently reflect existing State practices. Some of the upper riparian States were opposed to the "basin approach" adopted by the Helsinki Rules. States, therefore, requested the ILC to examine this issue in 1970. It took more than two decades for the ILC to finalize the draft framework treaty on the subject of law of non-navigational uses of international watercourses. In 1994, the ILC submitted its draft, which was adopted by the UN General Assembly in May 1997. Now, the 1997 UN Convention remains the authoritative source of law for the non-navigational aspects of river water sharing. Though this Treaty has come into force, none of the South Asian countries have joined this treaty thus far. Despite this, States usually accept that many of the provisions of the 1997 UN Convention have almost attained the status of customary norms of international law, such as the equitable apportionment of water and the general obligation to cooperate and notify co-riparian States in case certain planned measures are likely to cause significant harm.

Linking both the domestic and global strands of the study, the South Asian countries need to cooperate with each other to enhance the utilization of their water resources while applying emerging global norms. Existing bilateral treaty structures with regard to water sharing among the South Asian countries are certainly contributing towards the application of some of these basic principles and norms despite intermittently difficult political relationships. However, those efforts towards cooperation and sustenance of "good faith" obligations among the countries of South Asia should continue with the future endeavours.

Chapter 3
Agreements and Institutions Related to Shared Rivers within South Asia and Beyond

Shafqat Kakakhel

Six countries of South Asia—Afghanistan, Bangladesh, Bhutan, India, Nepal and Pakistan—share the water flows of twenty major and scores of smaller rivers. The three main river systems are the Indus, the Ganga or Ganges, and the Brahmaputra. The Indus Basin includes areas in China, Afghanistan, India, and Pakistan. The Brahmaputra and the Ganges connect China, Bhutan, India, Nepal, and Bangladesh. At present South Asia is arguably the only region without any regional mechanism or initiative on the sharing of the common trans-boundary rivers. Thus far, bilateral agreements, memoranda of understanding (MOU's), and treaties have been signed by India and Pakistan, India and Nepal, and India and Bangladesh on the sharing of rivers flows and utilization of water resources for irrigation and energy generation. Whilst the India–Pakistan Indus Waters Treaty (IWT), signed in 1960, divided six rivers of the Indus Basin between them on a permanent basis, the various agreements between India and Bangladesh, including the Ganga/Ganges Treaty of 1996, concern flows during the dry or lean period in respect of only one out of the fifty-four rivers flowing from India into Bangladesh. An agreement on another shared river—the Teesta—which was expected to be signed in September 2011 was effectively blocked by the spirited opposition of the West Bengal Government. The Indo–Nepal Treaty of 1996, which comprised revised versions of four earlier agreements (reached in 1920, 1959, 1966, and 1991), on the Kosi and Gandak rivers and added provisions for sharing of waters year round and construction of a big joint hydro-electric and irrigation project on the Mahakali river, has remained inadequately implemented due to political problems.

Afghanistan and Pakistan share the Kabul/Kunar River, a major tributary of the Indus which originates in Pakistan, crosses the border with Afghanistan to join the Kabul River which enters Pakistan and merges with Indus at Attock adding 18% to its flows. Both Afghanistan and Pakistan qualify as upper and lower riparian of this unique water course which contributes 12% of Afghanistan's total water resources and caters to the needs of over three million residents of Afghanistan's capital Kabul. Despite recognition of the need for enhanced cooperation and collaboration in the optimum development and utilization of the shared waters for mutual benefit

I. Ahmed (ed.), *South Asian Rivers*, The Anthropocene: Politik—Economics—Society—Science 21,
https://doi.org/10.1007/978-3-319-67374-5_3

highlighted during multi-stakeholder consultations held in 2009, officials of Afghanistan and Pakistan have not initiated formal discussions on the subject mainly due to turbulent conditions in Afghanistan since the mid-1970s.

Bhutan and India are jointly utilizing water resources of shared rivers for hydro power generation but do not seem to have felt the need for a formal water sharing agreement.

As pointed out in a recent survey carried out by Chatham House, "discussion about water in South Asia—in particular the shared rivers of the region—is vociferous, antagonistic and increasingly associated with national security". Tension over access to water within and between countries in the region is growing as water scarcity increases, and the discourse is increasingly focused on security. The implementation of each one of the water-related agreements in South Asia has been marked by differences over interpretation of key provisions and/or perceptions of willful violation by India widely perceived in the region as an unsympathetic hydro-hegemon. Given India's large size and economic and military prowess, its smaller neighbours seem to suffer from both "lower riparian" and smaller neighbour anxieties. India is upper riparian in respect of rivers shared with Pakistan and Bangladesh, where "lower riparian" concerns have been pronounced. The implementation of even the Mahakali Treaty on the utilization of the waters of the rivers for which India is the lower riparian has been hampered by opposition from many Nepalese political parties and water experts.

A water affluent region until recently, South Asia now faces the twin challenges of growing scarcity and deteriorating quality of freshwater resources. These challenges have been driven by rapid population growth accompanied by irregular, chaotic urbanization; industrialization and mining; intensive irrigation and agriculture; inefficient water use and governance; and decaying infrastructure. Per capita availability of freshwater in South Asia is said to have declined by 70% since 1950. Daunting in themselves, these woes are likely to be exacerbated by the multiple adverse impacts of climate change most of which relate to water, especially the rapid melting of the glaciers and disruption in the patterns of the monsoons which together feed the river flows and replenish the groundwater in the region.

Increasing awareness of the impending water crisis among South Asian stakeholders—especially experts, former diplomats, and civil society activists—has served to forge a widespread consensus on the need for enhanced cooperation and collaboration among South Asian states in the management of the region's water resources.

This paper reviews the background and key elements of the water-related agreements signed by India, Pakistan, Bangladesh, and Nepal, including institutional mechanisms for implementation and dispute settlement. It also refers to a few basin-wide agreements and initiatives in other parts of the world and concludes by making a plea for cooperation among South Asian countries for the optimum and equitable utilization of shared water resources in order to counter threats to their water, food, and energy security and help achieve sustainable and inclusive development and poverty eradication in South Asia.

3.1 Agreements on Shared Rivers Waters Negotiated by India and Nepal

Four river systems, each with its headwaters in the Tibetan Plateau, flow through Nepal into India. These are: the Sapta-Kosi, the Gandaki, the Karnali, and the Mahakali. These trans-boundary rivers contribute a large volume to the water resources of the entire region.

Despite geographic contiguity and a variety of affinities and the absence of any serious dispute, India and Nepal have not been able to fully cooperate in the development of water resources, perhaps because of the extreme sensitivities and divergent interests and approaches of the political parties in both countries. Water-related positions of both countries are based on political considerations rather than relevant technical and scientific factors. Whilst the Indo–Nepal Treaty of Peace and Friendship signed in 1950 was meant to strengthen friendly relations and provided for “equal and reciprocal treatment” of their citizens, a side letter which is an integral part of the Treaty laid down a number of restrictions on Nepal’s relations with third countries seems to have cast a dark shadow over their relationship.

Tragically, water-related agreements in South Asia have been counterproductive: they have provoked mistrust and reduced good will between India and Nepal. India is accused of having legitimized its uses of shared waters and unwillingness to give them up; it is being widely perceived as being indifferent to the difficulties and grievances of Nepal.[1]

3.1.1 *The Sarada Barrage Agreement*

The arrangements stipulated in exchange of letters between the British Indian Government and the Nepal Government in 1920 can be regarded as the first agreement between the two countries related to the Greater Ganges Basin. The agreement:

- Enabled India to build a barrage—the Sarada Barrage, also called the Banbasa Barrage—to divert water from Mahakali–Sarada for irrigation in the then United Provinces, now Uttar Pradesh;
- Transferred 4,000 acres on the eastern bank of the Mahakali in Nepal to India for building the barrage in return for an equal amount of land from India to Nepal further to the east plus a monetary compensation.
- Allowed Nepal to get an annual flow of at least 400 cusecs and maximum 1000 cusecs of water (out of the total annual flow of around 60 cusecs) while India’s withdrawals were not specified. Nepal’s lack of access to additional water is said

[1]See: Price, G. et al. (2014) Attitudes to Water in South Asia, Chatham House Report, Royal Institute of International Affairs.

to have made it impossible for Nepal to develop an upstream project. However, the agreement was operational until its replacement by 1996 Mahakali Treaty.

3.1.2 *The Koshi Project Agreement*

The Koshi Agreement was signed on 25th April 1954; a revised agreement was signed on 9th December 1966. Koshi is the largest river of Nepal. It originates in the Tibet plateau and along with its seven tributaries, it is the largest tributary of the Ganges. It is a wild river that used to cause floods annually in Bihar. In 1951 India developed a project for harnessing the river flows and reducing floods, producing electricity, and irrigating additional lands. After Nepal agreed in the 1954 Agreement, a barrage to 'discipline' the behaviour of the river and eastern and western embankments were completed in 1962 and 1982 respectively.

In the wake of deterioration of Indo–Nepali relations and criticism in Nepal of the negative fallout of the project on Nepal, the Indian Government agreed to extensively revise the terms of the agreement. The revised agreement was signed in 1966. The revisions of the agreement included:

- The establishment of a joint Indo–Nepal Kosi Project Commission to replace the Coordination Committee setup under the 1954 agreement for facilitating cooperation and coordination in the implementation of all matters covered in the agreement, including acquisition of land by Nepal for lease to India, rehabilitation of displaced persons, and maintenance of law and order.
- The amended agreement replaced 'ownership' of the land given to India by 'lease' and confirmed Nepal's sovereign rights and territorial jurisdiction over the land.
- The terms and conditions pertaining to the amounts of water and power to which Nepal was entitled were improved in the revised agreement.
- The Agreement recognized the navigation and fishing rights of Nepal.
- The Agreement's dispute settlement clauses provided for arbitration.

Nepal continues to complain that the Koshi agreement is more favourable to India, pointing out that while 356,310 ha of land are irrigated on the Indian side, only 11,300 ha are irrigated in Nepal. They also complain that the period of validity (199 years) is usually long; that there is no provision of a fresh revision even if considered necessary by Nepal; and that India has not maintained the barrage properly. Non-payment of compensation for the land 'lost' by Nepal is also resented. Issues such as "the incompletion of the Koshi project or the incompletion of the power house that was to be built (at) … Nepal's request" and inundation and displacement of people because of floods are additional reasons for "the breaching of the Koshi treaty"[2].

[2]Ibid, p. 25.

3.1.3 The Gandak Irrigation and Power Project Agreement

3.1.3.1 Key Elements

The agreement was signed on 4th December 1959 and allowed India to build at its own expense a barrage on the Gandak (Narayani in Nepal) at the Indo–Nepal border for reducing floods in India and diverting water into Bihar and Uttar Pradesh for irrigation and power generation. The project led to irrigating more than 1,850,520 ha of land in India and 4,700 ha in Nepal from the eastern canal, which passes through a few kmin Nepal before entering Indian Territory. Another canal—the Nepal Western Canal—covers a command area of 16,000 ha. "Actually the Gandapur project irrigates approximately 143000 acres in Nepal and 4.6 million acres in India".[3] The power house on the western canal in Nepal has the capacity to produce 15000 KW of electricity to be shared equally by India and Nepal. The agreement specified Nepal's share of waters but not India's. The agreement did not specify the time framework for the agreement, which meant that it would be valid forever!

The adverse consequences of the irrigation project for Nepal caused by the Gandak project were identical to those resulting from the Kosi Agreement, namely, inundation and loss of agricultural land and displacement of people. The alleged poor maintenance of the drainage system by India is said to have caused water logging in Nepal.

3.1.4 The Tanakpur Barrage Agreement

The Tanakpur Agreement was signed in the form of a Memorandum of Understanding (MOU) on 6th December 1991. The Agreement:

- Extended the scope of the Sarada project to include the Tanakpur Hydelproject started in 1988 on the same river.
- Allowed India to use 2.9 ha of Nepali territory for building an afflux bund for generating power mainly for India but allowing Nepal 10 MW. Unlike the Sarada Agreement, it did not provide for an even exchange of land.
- Required India to install power generation equipment on the barrage and build a canal to provide 150 cusecs of water for irrigating 4,000–5,000 ha of land.
- Recognized Nepal's navigational rights, subject to their being "technically feasible"—a condition which is said to have provided a pretext to India to defer the improvement of navigational facilities.

[3]See: Kishore Uprety and Salman MA Salman's essay entitled "Legal Aspects of Sharing and Management of Transboundary Waters in South Asia: Preventing Conflicts and Promoting Cooperation" in the publication Hydrological Sciences Journal, 2011.

The Tanakpur Agreement was criticized by most Nepali political parties for being discriminatory to Nepal and was not ratified. India refused to re-negotiate its terms.

3.1.5 The Treaty on the Integrated Development of the Mahakali River

The fifth Indo–Nepali agreement called "The Treaty Concerning the Integrated Development of the Mahakali River including the Sarada Barrage, the Tanakpur Barrage and the Pancheshwar Project" was signed on 12 February 1996. It was duly ratified by Nepal's parliament.

The preamble of the Treaty recognizes the Mahakali as a "boundary river on major stretches between the two countries", refers to the existing instruments (agreements) and decisions of the Joint Commission, and proclaims the intention of the two countries to jointly develop and implement a Detailed Project Report (DPR) on the development of the water resources of the Mahakali River.

The Mahakali Treaty comprises three parts. Part 1 (Article I) incorporates key provisions related to the Sarada Barrage providing for supply of 1000 cusecs of water from the Barrage to Nepal during the wet season and 150 cusecs during the dry season. India is to maintain a flow of not less than 350 cusecs downstream of the Barrage in order to preserve the ecosystem of the river.

Part 2 (Article II) of the Treaty relates to the Tanakpur Barrage and stipulates the works to be undertaken by India and Nepal in the Barrage area including the eastern afflux bund for which Nepal is to provide 2.9 ha of land which will remain under Nepal's sovereignty. Nepal is to receive 1000 cusecs of water in wet and 300 cusecs in dry season. Nepal is also to receive 70 million KW hours of electricity through transmission lines built by India.

Part 3 (Article III) deals with the Pancheshwar Multi-purpose Project upstream of the Tanakpur Barrage whose details were to be fleshed out in a Detailed Project Report (DPR) within six months of the conclusion of the Treaty. The Pancheshwar dam was to irrigate 93,000 ha of land in Nepal and 1.6 million hectares in India whilst the power house would generate 6500 MW of electricity. The project was also envisaged to help in preventing floods.

The Treaty laid down some specific principles on the sharing of the trans-boundary river, such as the principle of "equal rights and obligations" and "equal entitlement in the utilization of the waters of the Mahakali River". However, the principle of equal entitlement was qualified by a reference to "without prejudice" to their (India's) "existing consumptive uses" of the waters of the Mahakali river prior to the conclusion of the Treaty, particularly after the construction of the Pancheswar project. This amounted to undermining Nepal's rights in view of the non-existence of a clear definition of "consumptive uses". The linking of the "benefit sharing" with the costs to be incurred by the parties has also been viewed as being more favourable to India.

The Mahakali Treaty is valid for 75 years and is subject to review after 10 years and is to be amended of necessary.

A joint Indo–Nepal Mahakali River Commission comprising an equal number of representatives from both countries was mandated to serve as the main institutional mechanism for the development of the water resources of the Mahakali River. The expenses of the Commission are to be borne equally by both the Parties.

The dispute settlement mechanism enshrined in the Treaty represents an advance on the provisions of the earlier agreements. Each party is to designate an arbitrator. The two arbitrators would jointly select an 'umpire' failing which either Party may request the Secretary General of the Permanent Court of Arbitration to nominate an arbitrator to serve as 'umpire'.

The Mahakali Agreement having been ratified by both parties took legal effect in June 1997. However, the implementation of the Treaty has run into difficulties due to intense hostility in Nepal from politicians and experts and has remained in limbo. The Detailed Project Report prepared by the experts of the two countries has not been finalized. Officials of both countries have given divergent interpretations of the various provisions of the Treaty, especially the reference to "existing consumptive uses." However, "several agreements or contracts for hydropower projects in Nepal have subsequently been signed between Indian public and private companies and the Government of Nepal, applying the principle of build, own, operate and transfer (BOOT)." But even some of those agreements have been contentious and are perceived to be unfair to Nepal.

Prime Minister Narendra Modi's visit to Nepal on 3–4th August 2014—the first such visit after a hiatus of 17 years—has raised hopes of a major positive turn in Indo–Nepali relations and of the implementation of the Mahakali Treaty. Significantly, the joint statement on the outcome of the Prime Minster Modi's visit mentions an agreement "to review, adjust, and update the Treaty of Friendship of 1950 and other bilateral agreements." During the visit, an exchange of letters was signed spelling out the Terms of Reference (TOR's) of the Pancheshwar Development Authority (PDA), which will formalize the Detailed Project Report provided for the Mahakali Treaty, signalling better prospects for water resource-centered cooperation.

3.2 Agreements Between Bangladesh and India

India had played a decisive role in the liberation of Bangladesh in 1971. The Treaty of Friendship, Cooperation, and Peace signed on 19th March 1972 had proclaimed the shared goals of friendship and cooperation in various fields, including "flood control, river basin development and development of hydroelectric power". However, Indo–Bangladesh relations have oscillated between periods of cordiality and tension for a number of reasons. Non-resolution of old and emergence of new disputes over shared water resources have been both a cause and consequence of the un-satisfactory state of bilateral relations between the two neighbours.

Bangladesh and India share fifty-four rivers some of which such as Kushiara, Mutturi, Feni, Ishmati, and Gumti pass through their border marking their territorial boundaries. A land boundary agreement signed in 1974 provided for the demarcation of the boundaries along the mid-streams of the river courses. So far, only one comprehensive agreement—the Ganges Treaty of 1996—has been signed by the two co-riparians and its implementation has neither been smooth nor trouble-free.

Of the three large river basins shared by Bangladesh with its neighbours, the Ganges has been the most contentious. The Ganges rises in the slopes of the Himalayas at the border between Nepal and China and between India and China. The tributaries originating in Nepal and China, including the Kosi, Gandaki, Karnali, and Mahakali, supply near half of the Ganges flows underlining the critical need for cooperation among Bangladesh, China, India, and Nepal.

At its independence, Bangladesh inherited the decade old India–Pakistan dispute over the construction by India of a barrage over the Ganges at Farakka—known as the Farakka Barrage—about 10 km from the border. Pakistan's efforts to forestall the building of the barrage through bilateral means and involvement of international organizations were thwarted by India. The 240-m long barrage was completed in 1971; the feeder canal for delivering water from the dry season (January–May) flows of the Ganges to the Hoogly River became operational in 1975.

India has consistently argued that the Farakka Barrage was needed to make the Hoogly navigable in order to make the Calcutta (now Kolkata) port accessible by flushing down silt from the port as well as provide water for domestic use, irrigation, and other needs. The Governments of Pakistan (until 1971) and Bangladesh (since 1971) have contended that dry season flows were indispensable for irrigation, domestic consumption, hydropower generation, fisheries, and forestry on its side of the divide.

The fact that the problem of Farakka barrage was raised by the Father of Bangladesh, Sheikh Mujibur Rahman, during his first visit to India in February 1972 underscores the importance of the subject for Bangladesh. The issue also figured in the joint declaration on the establishment of the Joint River Commission (JRC) in March 1972. The JRC was meant to enable the two countries to work together "in harnessing the rivers common to both countries for benefit of people of the two countries" and to "maintain liaison between participating countries in order to ensure the most effective joint efforts in maximizing the benefits from the common river systems to both countries". The Commission was also entrusted to formulate and implement flood control measures. The statute of the JRC was signed in November 1972 according to which it was to meet "as often as necessary, generally four times a year".

By May 1974, Bangladesh seems to have reconciled to the Farakka Barrage as a "fait accompli" during the third visit of Sheikh Mujib to India. The joint declaration on the visit proclaimed the agreement of the two countries on the imperative of augmenting the dry season flows of the Ganges in order to meet the needs of both countries. Bangladesh conceded that the augmentation would be through the "optimum utilization of the water resources of the region available to the two countries"—an agreement that "would come to haunt Bangladesh when the discussion on augmenting the flows of the Ganges started.[4]" Bangladesh had inadvertently accepted India's position regarding diverting water from the Brahmaputra in order to augment the flow of the Ganges!

At the JRC, Bangladesh suggested a number of schemes for storing the monsoon flows of the Ganges in India and Nepal to augment the flows during the dry/lean months. India came up with the counter proposal of transferring water from the Brahmaputra to the Ganges. This was turned down by Bangladesh on several technical grounds.

3.2.1 Agreements on the Sharing of the Ganges Flows

The first 'agreement' on the Ganga/Ganges was reached in April 1975 and called the "Partial Accord" in a press release whereby Bangladesh agreed to the test running of the just completed feeder canal of the Farakka Barrage. The 'agreement' expired on 31st May 1975 but India continued to withdraw water to the full capacity of the feeder canal.

In 1976 amid deteriorating ties between Bangladesh and India following the assassination of Sheikh Mujib, Bangladesh complained to the UN against the Farakka Barrage. This triggered a bitter diplomatic row in which both sides disseminated documents substantiating their respective positions on the dispute. The UN General Assembly called for a fair and expeditious settlement of the dispute through negotiations. This and the installation of a new, non-Congress government in New Delhi headed by Morarji Desai facilitated a series of ministerial-level discussions culminating in the signing of the second agreement on 5th November 1977 on sharing of the dry season (1st January–31st May) flows of the Ganges available at Farakka for a period of 5 years (1978–1982).

The "Agreement between the Government of the Republic of India and the People Republic of Bangladesh on Sharing of the Ganga Waters at Farakka and on Augmenting its Flows" comprises fifteen articles related to the sharing of flows during the dry months and the need for augmenting the flows. The schedule of flows is contained in Annex 1 of the agreement allocated at Farakka to Bangladesh. Agreement provided that in case of especially low flows at Farakka, Bangladesh would be guaranteed at least 80% of its share for 10 day segments. This became known as the Guarantee Clause of the 1977 Agreement.

The 1977 Agreement entrusted the JRC to submit jointly evolved recommendations on the augmentation to the two governments within three years. During the JRC meetings, both countries reiterated their well-known positions. Eight meetings

[4]Salman, M.A. and Uprety, K. (2003). Conflict and Cooperation on South Asia's International Rivers: A Legal Perspective, Washington DC.

of the JRC held during the period between January 1978 and September 1980 failed to agree a solution.

The 1977 Agreement expired on 31st May, 1982. India turned down Bangladesh's proposal for extension of the agreement as provided for in Article XV of the agreement.

The third agreement between Bangladesh and India was signed as a Memorandum of Understanding (MOU) on 7th October 1982 for sharing the flows of the Ganges during the dry seasons of 1983 and 1984. The MOU also stipulated an 18-month deadline for completing the pre-feasibility study on augmenting the dry season flows at Farakka.

During the subsequent meetings of the JRC, Bangladesh proposed seven storage reservoirs in Nepal for harnessing the Ganges flows. Once more, India suggested transferring 100,000 cusecs of dry season Brahmaputra flows to Farakka through a 200-mile Brahmaputra—Ganges Link Canal. Following protracted but in fructuous discussions, the JRC on 3rd March 1984 admitted having failed to develop a solution acceptable to both countries. A deadlock ensued, leading to reduced flows to Bangladesh.

The fourth Indo–Bangladesh agreement was signed as yet another MOU on 22nd November 1985 in the wake of meetings between the President of Bangladesh and the Prime Minister of India. The new MOU called for studies by a joint committee of experts (JCE) on a long-term scheme or schemes for augmentation of the flows of the Ganges at Farakka within a period of twelve months. The JCE discussed the proposals submitted by the two sides for twenty four months but without success. The sharing arrangements under the 1985 MOU expired on 31st May 1985. In the absence of an agreement, India again reduced the flows to Bangladesh. Five sessions of talks between water secretaries of the two countries held between June 1990 and August 1991 could not make any breakthrough.

Meetings between the Prime Ministers of Bangladesh and India in May 1992, April 1993, and May 1995 and talks at Foreign Secretary level failed to break the stalemate. The formation of an Awami league government under Sheikh Hasina Wajed facilitated resumption of negotiations, which led to the signing of the Ganges Treaty on 12th December 1996 in New Delhi during the visit of Prime Minister Sheikh Hasina Wajed.

3.2.2 *The Treaty on the Sharing of the Ganga/Ganges Waters at Farakka (12 December 1996)*

In terms of its main elements, the Treaty:

- Is valid for 30 years (unlike previous agreements of much shorter periods)
- Deals with the quantum of waters agreed to be released by India to Bangladesh. The quantities are prescribed in Annexes I and II.

- Specified that the amounts of waters to be shared relate to 10-day periods during the dry/lean months (January–May). The amounts have been calculated on the basis of flows during a 40-year period (1949–1988) unlike previous agreements whereby releases were calculated on the basis of 75% of availability of the average flows during the years 1948–1975.
- Stipulates that India's total share during the lean period is around 43% of the available flows; Bangladesh's share represents around 52%. During the critical periods of exceptionally low flows each country is to receive 35,000 cusecs of waters. In case the flow at Farakka is less than 50,000 cusecs, the signatories are to hold "immediate consultations" in order to make adjustments on an emergency basis, in accordance with the principles of equity, fair play, and no harm to either party.
- States recognition by both countries of "the need to cooperate with each other in finding a solution to the long term problem of augmenting the flows of the Ganga/Ganges during the dry season."
- States the agreement of both countries "to conclude water—sharing Treaties/Agreements with regard to other common rivers."

The Ganges Treaty has been both lauded and criticized. Favourable comments mention nomenclature (Treaty signifies stronger commitment than an agreement or MOU); longer period of validity; recognition of the rights of the lower riparian; and references to the "the principles of equity, fair play and no harm to either party."

Critics describe the Treaty as "a water sharing agreement in its most primitive sense: it is solely an arrangement for the volumetric allocation of river flow in the dry season" which "does not concern benefit sharing, nor does it purport to be a comprehensive river sharing and management treaty".[5] The calculation of allocations on the basis of the 1949–1988 flows means no relief for Bangladesh from the decrease in the dry discharge of the Ganges at Farakka "due to increased upstream uses for agriculture and other purposes". The Treaty is said to favour the hydro-hegemonic state (India, "solidifies the status quo" and has left Bangladesh with "numerous concerns and unresolved issues"[6]).

Environmental experts have corroborated the several serious adverse impacts of the Farakka Barrage such as reduced flows to Bangladesh, intrusion of saline waters during the dry months which have damaged the Sundarban forests, agriculture, and fisheries forcing large scale migration of communities from Bangladesh to India that has exacerbated tensions in the north-eastern and other parts of India.

The omission of any guarantees of minimum flows in the Treaty, unlike those stipulated in the previous agreements, has especially hurt Bangladesh. In fact, during the lean period in 1997—a few months after the signing of the Treaty—the availability of water was much less than 50,000 cusecs. Bangladesh reportedly received only 24,559 cusecs, instead of the 39,106 cusecs as stipulated in the

[5]Hanasz, P. (2014). "Sharing Waters vs. Sharing Rivers: The 1996 Ganges Treaty", *Transboundary*, July. Global Water Forum.

[6]Ibid.

Treaty. Consultations sought by Bangladesh under the Treaty did take place but no adjustments were made.

There has been no headway in finding a solution of the most crucial question of augmentation of dry season flows despite numerous meetings at technical and political levels.

Bangladesh experts have complained about receiving reduced flows at Farakka with all the associated negative consequences. In May 2007, the India's co-chair of the Joint Committee entrusted with the implementation of the Ganges Treaty is said to have declared India's intention to divert 40,000 cusecs of water and release the rest to Bangladesh regardless of the actual availability of water. He is also said to have linked water availability to adequate rainfall in the upstream as well as attributed reduced flows in the dry season to the impacts of climate change.

3.2.3 Bangladesh–India Dispute over the Teesta and Other Shared Rivers

Another longstanding shared rivers-related dispute between Bangladesh and India concerns the Teesta River, which originates in the north east Indian state of Sikkim and flows through parts of West Bengal before entering Bangladesh through 45 km^2 of irrigated land and merging with Brahmaputra. The livelihoods of over 9 million people in five districts of Rangpur Division depend on the Teesta flows.

Both Bangladesh and India have built or planned barrages and link canals in their tributaries to expand irrigation. Their dispute is over the release of water by India during the dry/lean months when flows dwindle.

In 1983, protracted negotiations on the Teesta River initiated in 1972 under the ambit of the JRC led to an interim agreement allocating 39% of the dry season flows to India and 36% to Bangladesh, leaving the rest un-earmarked pending a permanent solution. The agreement expired in 1987. Subsequently the subject was taken up by a Joint Committee of Experts (1997–2004) and a Joint Technical Group (2004–2010) but without success.

The Awami League's electoral victory in December 2008 energized the quest for a settlement and a draft agreement and a set of guiding principles were prepared. The signing of the agreement was expected to take place in September 2011 during the visit of Prime Minister Manmohan Singh to Dhaka but was scuttled at the last minute by the fierce opposition of the West Bengal Chief Minister Mamata Banerjee. Mrs. Banerjee protested the lack of prior consultation with her government as well as the quantity of water allocated to Bangladesh in the proposed deal.

In pursuance of the agreement of the two countries "to conclude water-sharing Treaties/Agreements with regard to other common rivers" stipulated in the Ganges Treaty, a committee of experts constituted by the JRC has identified seven common rivers which warrant cooperative arrangements for sharing, namely Muhuri, Manu, Gumti, Khowai, Brahmaputra, Dharla, and Dudh Kumar.

Bangladesh and India signed a "Framework Agreement on Cooperation for Development" in September 2011 during the visit to Dhaka by the Indian Prime Minster. Article II of the Agreement stated: "To enhance cooperation in sharing of the waters of common rivers, both Parties will explore the possibilities of common basin management of common rivers for mutual benefits." A Joint Consultative Commission (JCC) was established for "effective and smooth implementation" of the Framework Agreement which was to meet once a year. The JCC has met regularly since 2012.

President Pranab Mukerjee assured Bangladesh's leaders during his visit to Dhaka in March 2013 that India was committed to "a fair, reasonable solution" of the Teesta issue. However, during the dry season of 2014 India reportedly released only 10% of the flows to Bangladesh.

The passage of the 100th Constitutional Act by the two houses of the Indian parliament in May 2015 paved the way for the ratification and operationalization of the 1974 Land Boundary Agreement providing for exchange of territories between Bangladesh and India covering more than 150 enclaves along Indo–Bangladesh borders inhabited by tens of thousands of persons whose national status had remained un-determined for decades.

Prime Minister Narendra Modi visited Dhaka on 6–7 June 2015. He was accompanied by a delegation that included leading Indian industrialists and entrepreneurs, and was joined by Chief Minister Mamata Banerjee of West Bengal who had opposed the signing of the Teesta river agreement. On 6 June the formal exchange of Instruments of Ratification of the Land Boundary Agreement took place. The outcomes of discussions before and during the visit on multiple areas of economic, technical, and technological cooperation were catalogued in a 65-Para Joint Declaration issued on 7 June. Paragraphs 18–22 of the Declaration address shared river water issues. The key points are noted below:

Ganges Bridge: Indian agencies will 'examine' Bangladesh's request for cooperation in jointly developing the Ganges (also called Padma) Bridge on the Padma River.[7]

Teesta and Feni: In response to Prime Minister Sheikh Hasina Wajed's request for "the immediate conclusion of the Interim Accord on the sharing of the Water of the Teesta River", Prime Minister Modi referred to the "deliberations (that) are underway involving all stakeholders" on the "conclusion of the Agreements on Sharing of Waters of Teesta and Feni as soon as possible". Prime Minister Modi is reported to have promised "to make it (the Teesta Agreement) happen".

Agreements on other Shared Rivers: The two Prime Ministers 'noted' the discussions under the JRC on various agreements on the sharing of the waters of Manu, Muhuri, Khowai, Gumti, Dharla, and Dukh kumar rivers and asked the concerned officials to take expeditious action to conclude the sharing arrangements "at the earliest".

[7]The bridge will connect the south west regions of Bangladesh with the northern and eastern regions.

Cooperation in basin-wide management of common rivers: The Prime Minister recalled Article II of the Framework Agreement on Cooperation for Development of 2011 and "reiterated their commitment to address the issue of water resource management of common rivers including water sharing in a holistic manner through common basin management".

The Indian Rivers Inter-linking Project: Prime Minister Modi "reiterated the earlier commitment that India would not take any unilateral development on the Himalayan component" of the Project "which may affect Bangladesh".

Tipaimukh Hydro-Electric Project: Prime Minister Modi mentioned "statuary requirements in India" holding up the project and added that "India would not take any decision on this matter that may adversely impact Bangladesh".

JRC: The Prime Minister "stressed the need for holding the 38th ministerial meeting of the JRC at an early date to discuss all issues pertinent to water resources for mutual benefit (the 37th meeting of the JRC had been held in March 2010!)".

Based on the wording of the Joint Declaration, Prime Minister Modi's visit does not seem to have led to any concrete decision on the issues related to shared transboundary rivers.

3.2.4 The India–Pakistan Agreement on the Indus River Basin: The Indus Waters Treaty (IWT), 1960

The Indus Basin (IB) comprises six major and twenty seven smaller rivers. The Indus itself originates in the Tibetan Plateau, flows through the Ladakh district of Jammu and Kashmir, and then enters Pakistan where it flows through parts of all four provinces of the country before entering the Arabian Sea near Karachi. The Sutlej also rises in Tibet and flows through India before entering Pakistan. The other four tributaries—the Jhelum, the Chenab, the Ravi and Beas-rise in India and flow through Pakistan. The Indus Basin covers an area of 1.2 million km^2 of which 47% lies in Pakistan, 39% in India, 8% in China, and 6% in Afghanistan. More than 300 million people in India, Pakistan, and Afghanistan depend on the waters of the Indus Basin for domestic consumption, irrigation, and other uses. The IB is the sole river system in Pakistan and its crucial mainstay given its indispensable role in sustaining agriculture which contributes a fifth of the country's GDC, 45% of all jobs, livelihoods of the majority rural population and produces raw material for Pakistan's predominantly agrarian economy.

Since mid-19th century, the British Raj had constructed an extensive canal irrigation network in the Punjab for commercial agriculture turning it into one of the major food baskets of the sub-continent. The partition of India in 1947 disrupted the inter-connected irrigation system; the upstream of the rivers and the head works of the irrigation canals went to India although 80% of the irrigated lands of Indus Basin were in territories that formed part of Pakistan.

Efforts to develop post-partition water sharing arrangements on the Indus Basin Rivers prior to August 1947 failed. A standstill Agreement ensured water flows until the end of March 1948. Sudden, un-notified termination of water supply by the (Indian) Punjab state government on 1st April at the start of the summer (irrigation) season affecting 5.5% of Pakistan' irrigated area triggered the breaking out of a major water dispute between India and Pakistan. Water supplies were subsequently restored and an Inter-Dominion Agreement was signed on 4th May 1948. However, the agreement became controversial and tension persisted. India rejected Pakistan's proposal to refer the dispute to the International Court of Justice for settlement.

In 1951, the World Bank agreed to serve as a "go-between", a "post office" and "an impartial adviser... but not an arbitrator" in discussions between Indian and Pakistani engineers aimed at developing "a comprehensive long-range plan" for "the cooperative development and use" of the waters of the Indus Basin "viewed as a unit". In the process, the Bank played a "pivotal role" in facilitating agreement on the key elements of what became the Indus Water Treaty (IWT) signed by the Prime Ministers of India and Pakistan on 19th September 1960. World Bank President Eugene Black treated the Indo–Pak negotiations as a top priority and carried out shuttle diplomacy visiting the two countries frequently to nudge the leaders toward compromises on core issues.

3.2.4.1 Key Elements of the IWT

The IWT documents consist of a short treaty comprising a preamble and 12 articles and eight long annexures (A–H). The key elements of the Treaty are noted below.

- The three eastern rivers of the Indus Basin—the Sutlej, Ravi, and the Beas—were allowed for the unrestricted use of India whilst the three western rivers—the Indus, Jhelum and Chenab—were allocated to Pakistan. India was allowed to use the flows of the western rivers for domestic consumption, irrigation and hydropower generation (run-of-the-river projects) under strict limitations stipulated in the long Annexures "C" and "D".
- Water allocations were not specifically mentioned in cusecs. However, based on the average flows, India's share came to 33 Million Acre Feet (MAF) of annual flows of the Eastern Rivers, whilst Pakistan was to receive the flows of the western rivers estimated at 165 MAF (or around 83% of the Basin's available flows).
- Going beyond the partitioning of flows, the IWT provided for the 'development' of the water assets through an array of infrastructural works to store and transfer flows of the western rivers to areas hither to irrigate by the eastern rivers. These included two large dams—at Tarbela on the Indus and at Mangla on the Jhelum—several barrages, eight link canals and over 3000 tube wells.
- In an unprecedented contribution, and with a view to enticing Pakistan to accept the loss of the vital water resources of the three eastern rivers, the World Bank established and actively mobilized funds for an Indus Development Fund, which

received over $1 billion from US, UK, Canada, West Germany, Australia, and New Zealand, as well as India (which provided $174 million in several instalments) for financing the construction of the engineering works.
- The IWT provided for an Indo–Pak Permanent Indus Commission (PIC) comprising a commission for Indus Waters setup by each Party, to serve as "the regular channel of communication on all matters pertaining to the implementation of the Treaty".
- The IWT provided for an elaborate, 3-tier dispute avoidance and settlement mechanism. The PIC serves as the first tier of the mechanism, charged with providing timely information on actions relevant to the implementation of the Treaty, and clarifying any technical 'questions.' In case the 'questions' are not sorted out, the World Bank is to appoint a "neutral expert" acceptable to both Parties to resolve the 'differences', whose finding is binding. In the event the neutral expert or either Party considers the 'differences' to be a 'dispute', the International Permanent Court of Arbitration (PCA) can be approached to arbitrate in the matter. The award/verdict of the PCA is binding on both Parties.
- The IWT provides for "exchange of data" in respect of flows in, and utilization of, the waters of the rivers and withdrawals from the canals, as well as any data relating to the hydrology of the rivers, canals, or reservoirs, sought by either Party, provided the data is available.
- The IWT also provides for "future cooperation" between India and Pakistan in the optimum utilization of the rivers, including engineering works, provided they do not interfere with the waters of the rivers.
- The IWT stipulated transitional arrangements regarding the continued supply of waters of the eastern rivers to Pakistan for a period of 10 years (whilst the civil works were being undertaken) until 31st March 1970.

3.2.4.2 Implementation of IWT

The implementation of the transitional arrangements and other provisions of the IWT proceeded smoothly until the mid-1970s, including during the 1965 Indo–Pakistan war. Subsequently, however, differences and disputes have arisen over the construction by India of a large number (over 35) hydro-electric projects on the Jhelum and the Chenab.

Of the four major issues related to Indian projects the first, over the building of a hydropower project on the Chenab, known as the Salal Dam, which arose in 1974 was resolved in April 1978 thanks to India's agreement to carry out design changes sought by Pakistan. The dispute was resolved through talks at the level of Foreign Secretaries.

The second dispute over the construction of a barrage on the Jhelum at the mouth of the Wullar Lake, referred to as the Tulbul Navigation Project by India and Wullar Barrage by Pakistan, arose in 1984 and has not yet been resolved. India contends that the project is meant to not only generate 450 MW of electricity but also regulate water flows in the river in order to facilitate year-round navigation linking several towns in Kashmir. Pakistan claims that the project would create storage beyond the limit set in the IWT.

The third dispute relates to the 450 MW hydro-electric project on the Chenab at Baglihar. Pakistan had objected to the design of the dam because it included low-gated spillways not allowed by the IWT. India contended that the gated spillways were needed to prevent the silting of the reservoirs as had happened in Salal Dam. A neutral expert appointed by the World Bank upheld India's position but suggested some changes in the design of the project to assuage Pakistan's concerns.

The fourth dispute relates to the complex 900 MW hydro-electric project planned by India for utilizing the waters of the Kishenganga River, a tributary of the Jhelum, to be diverted through a tunnel to another tributary of the Jhelum. Pakistan objected to the project claiming that the inter-tributary transfer of flows did not conform to the IWT. The Permanent Court of Arbitration (PCA) to which the dispute was referred in 2010 issued two verdicts in 2013 allowing India to precede with the construction of the project works but required it to increase 'environmental' flows to Pakistan. The PCA also ruled out low-gated spillways upheld by the neutral expert in respect of the Baglihar Project in future projects if undertaken by India on the western waters.

Pakistan has raised technical objections to several Indian hydropower projects which are currently under consideration by the Indus Commission. Pakistan's main fear is that the dozens of projects built by India on the western rivers whilst not being in violation of the IWT would nevertheless enable India to control and manipulate flows into Pakistan especially during the dry season which could be used to seriously harm Pakistan's agriculture and food security. Indian officials dismiss these concerns as symptoms of lower riparian anxiety and mutual distrust.

3.2.4.3 Future Prospects

The IWT has been widely and accurately hailed as a spectacular triumph of 'hydro-diplomacy' especially given the distrust, rivalry and hostility marking Indo–Pak relations and the un-resolved territorial/political dispute over Jammu and Kashmir. The World Bank is eloquently praised for its indispensable role in the negotiation and implementation of the Treaty. Notably, although negotiated three a half decades prior to the adoption of the landmark UN Convention on the Law of the Non-Navigational Uses of International Water Courses (UNCIW) in 1997, the IWT includes elements which could be construed as precursors of some of the key principles enunciated in the Convention such as those upholding equitable utilization; no harm to a co-riparian; prior and timely notification of actions likely to affect a co-riparian or co-riparians; peaceful, negotiated resolution of disputes; cooperation in the optimum development and utilization of waters; prevention of pollution etc.

Angry outbursts and criticism occasionally voiced by extremist elements pursuing political agendas unrelated to water in both India and Pakistan notwithstanding, the IWT is considered sacrosanct by governments, experts, and the broader civil society in both countries and beyond South Asia. Motivated by concern over the rapid decline and deterioration of freshwater which are likely to be heightened by the impacts of climate change, and eager to turn a potential flash point of conflict into a catalyst for mutually beneficial cooperation, dozens of retired officials, including former diplomats, experts and civil society representatives of India and Pakistan, joined by Kashmiri experts, have carried out informal discussions popularly called Track II dialogues, on water issues during the past decade. The focus of the discussions has been on identifying key questions concerning shared waters, which need to be examined in an objective, scientific manner jointly by the relevant institutions and experts of the two countries. It is hoped that findings of the studies based on impeccable science and empirical evidence would serve as a basis of informed decision-making on cooperation in water resource management by the Indian and Pakistani governments.

The questions identified by the Track II dialogues for joint studies include:

- Strengthening the capacity of the PIC to play the role prescribed for it by the IWT, including the provisions on the exchange of data and future cooperation (Articles 6 and 7) whose implementation has been patchy or non-existent.
- The impacts of climate change, especially monitoring and evaluating the state of the glaciers in the Himalayas serving as water stores of the Indus Basin and other rivers originating in the Tibetan plateau, and the patterns of the monsoons.
- Preserving the sustainability of the ground water resources, which are an integral part of river hydrology, given the excessive withdrawals of ground waters in both India and Pakistan for intensive irrigation and human consumption.
- An enhanced understanding of the cumulative capacity acquired by India from the cascades of Hydel projects on the Jhelum and Chenab.
- Protection of the upstream watershed of the western rivers aimed at curbing the unabated pollution of the flows by industrial and agricultural runoff and de-forestation in India and Jammu and Kashmir.
- Minimum environmental flows, including in the eastern rivers necessary for protecting the eco-systems.

Thus far Indo–Pakistan discussions on water issues have been limited to disputes over Indian hydel projects on the western ('Pakistani') rivers. A significant breakthrough in initiating discussing broader cooperation would require "political will" and resumption of the perennially hiatus-prone official dialogue between the two countries.

3.2.5 *Agreements and Institutions Related to Shared Trans-Boundary Rivers Beyond South Asia*

There are 263 international river basins which account for nearly one half of the earth's landmass, generate 60% of the global freshwater flows, and meet the needs of 40% of the world population. Whilst differences and disputes over water resources have frequently arisen in all parts of the world, thus far there has been no "water war" and co-riparians have been able to negotiate agreements and arrangements of varying scope and effectiveness on shared water resources.

Since the dawn of the 20th century, efforts to "preempt potential conflict and resolve existing disputes" have led to the evolution of principles of International law on the cooperative management of shared water courses.

Significant milestones in the development of International Trans-boundary Water Law include: The Madrid Declaration of 1911; the General Convention Relating to the Development of Hydraulic Power Affecting more than One State of 1923; the Swedish–Norwegian Convention of 1929; the agreement between Brazil and Uruguay in 1933; the US–Canada Boundary Water Treaty of 1909; the Helsinki Rules of 1966 on the Uses of the Waters of International Rivers; the Commission appointed by the UN in 1970 to codify law on the non-navigational uses of international water courses which culminated in the adoption by the UN General Assembly of the UN Convention on the Law of the Non-Navigational Uses of International Water Courses (UN Watercourse Convention) on 21 May 1997. The UN Convention broadly defined the principles of "equitable and reasonable utilization" and the "obligation not to cause significant harm" and established a framework for the exchange of data and information, the protection and preservation of shared water bodies, the creation of joint management mechanisms, the settlement of disputes, and measures to protect the ecological integrity of the watercourses.

The UN Watercourse Convention took legal effect in August 2014 after it had been ratified by 35 member states. Delay in the coming into force of the UN Convention and its non-binding nature notwithstanding, the principles enunciated and codified by it have contributed to the negotiation of agreements and institutional arrangements for cooperation among co-riparians in the optimum development and utilization of trans-boundary water resources.

Since 1948, approximately 300 international water agreements/treaties have been negotiated and signed which address a range of issues such as the quantity and quality of water, irrigation, and hydropower generation. More than two-dozen basin-wide agreements provide for the establishment of institutional mechanisms to ensure their implementation and further development. UN Water comprising over twenty agencies, the international financial institutions such as the World Bank and the Asian Development Bank, and forums and institutions devoted to promoting regional economic cooperation and integration such as the UN regional economic and social commissions; the European Union, the African Union, ASEAN, and the ministerial bodies in Latin America and the Caribbean have supported the implementation of the basin-wide agreements.

A few basin wide cooperation treaties in the developing world are noted below:

3.2.5.1 The Mekong Commission

The Mekong Commission was established in pursuance of the "Statement on Cooperation for the Sustainable Development of the Mekong River" signed by the four Lower Mekong riparian states (Cambodia, Laos, Thailand, and Vietnam) in 1995 for the development of the lower Basin through dams for irrigation and hydropower generation. The Commission serves as an inter-governmental mechanism for promoting "regional cooperation, facilitate dialogue on the sustainable management of the Mekong Basin's water resources" and avoid disputes. The Commission's programs relate to agriculture and irrigation; basin development; climate change and adaptation initiatives; drought management; environment; fisheries; flood management and mitigation; hydropower development, navigation; capacity building etc. Efforts are underway to cement cooperation between the Mekong Commission signatories and the Mekong's upper riparians, China and Myanmar.

3.2.5.2 The Nile Basin Initiative

The Nile is the world's longest river, travelling 6,700 km from Rwanda and Burundi through Lake Victoria to its delta in Egypt, covering 3.25 million km^2 representing 10% of Africa's total land area. The Nile serves more than 300 million people in 10 countries (Burundi, Democratic Republic of Congo, Egypt, Ethiopia, Kenya, Rwanda, South Sudan, Tanzania, and Uganda).

The Nile Treaty of 1929 between Egypt and Sudan (when both countries were under British control) was abrogated by Sudan in 1956 and a revised agreement was signed in 1959. The new Treaty allocated 18.5 billion m^3 of water to Sudan and 55.5 billion m^3 to Egypt on the basis of "prior appropriation." The shares of the other upper riparians were not specified.

The upper riparian states resent the terms of the Nile Treaty, which they allege have impeded the utilization of the Nile's waters for irrigation and hydropower by them because of the flows allocated to Egypt in the 1959 Treaty. However, in recent years considerable progress has been made in promoting dialogue and cooperation among co-riparians, especially with the Nile Basin Initiative launched on 22nd February 1999 as a transitional initiative until the finalization of the Cooperative Framework Agreement (CFA).

The objectives of the Nile Initiative are: (i) to develop the Nile Basin water resources in a sustainable and cooperative way to ensure prosperity, security, and peace for all its peoples, (ii) ensure efficient water management and optimum use of the resources, (iii) ensure cooperation and joint action between riparian states seeking win-win gains; (iv) to target poverty eradication and promote regional integration; and (v) transition from planning to action.

The Nile Initiative's activities are financed from a Nile Basin Trust Fund managed by the World Bank. The Fund's donors are Canada, Denmark, EU, Finland, France, the Netherlands, Norway, Sweden, UK, and the World Bank.

3.2.5.3 The Amazon Cooperation Treaty Organization (ACTO)

The Amazon River is the second longest river of the world; its waters represent one fifth of the planet's freshwater resources. Most of the basin formed by the Amazon River and its tributaries is covered by rainforests and is shared by 8 Latin American states (Bolivia, Brazil, Colombia, Ecuador, Guyana, Peru, Surinam, and Venezuela). The basin covers 6,915 km^2 area or roughly 40% of the South American continent.

The Amazon Cooperation Treaty was adopted on 3rd July 1978 and came into effect on 2nd August 1980; it was amended by a Protocol adopted in December 1998. The Amazon Cooperation Treaty Organization (ACTO) was setup in 1995 as an inter-governmental mechanism for promoting the sustainable development of the Amazon Basin. The Strategic Agenda for the Amazon approved by the member states of ACTO in November 2010 comprises programs and projects related to cooperation among the co-riparians in forests, health, education, transportation, hydropower, tourism, and science and technology. ACTO is also focusing on addressing the challenge of growing pollution of the Amazon Basin by industrial, agricultural and municipal run-off.

3.3 Recommendations and Conclusions

India, Pakistan, Nepal, and Bangladesh have concluded agreements on the trans-boundary shared rivers despite formidable challenges such as "stark political asymmetries, powerful strategic overtones, and deeply conflictive geo-political histories" as well as "the states' reluctance to budge from essentially self-serving attitudes and policies". Water issues are inextricably linked to the state of political relations and lack of trust among all four countries. As this paper notes, none of the agreements has either resolved water sharing disputes or spurred broader cooperation in the optimum development and utilization of the water resources on the basis of mutual benefit. The existing agreements need to be revisited and where necessary and possible revised in order to address new challenges and remedy institutional gaps and weaknesses. The dispute avoidance and settlement provisions of the Mahakali and the Ganges agreements need to be strengthened by adding a robust system as exists in the Indus Waters Treaty.

South Asian co-riparians are increasingly aware that their once water rich countries are headed for a future of growing scarcity thanks to population growth, development, and increasingly, the water-related effects of climate change which threaten their interlinked water, food, energy, and health security. Therefore, they need to enhance cooperation at bilateral and regional levels for basin-wide integrated management of shared water resources.

Co-riparian states need to recognize the pluralities of stakeholders in the governance and management of shared river flows and the imperative of facilitating the participation of non-state stakeholders, especially the civil society, in addressing the growing array of water-related issues.

South Asian co-riparians of shared rivers need to establish cooperative arrangements capable of responding to the unique conditions of their basin and imperatives of political economy. They should consider the useful lessons offered by successful basin-wide cooperative regimes forged by co-riparians in other parts of the developing world.

South Asian states need to revisit their reservations concerning the UN Watercourse Convention (1997). They also need to consider entering into a discussion on the desirability of a South Asian Declaration on the principles of basin-wide cooperation and take steps to implement the agreed principles in respect of transboundary watercourses shared by them.

Indian and Pakistani governments should take up the rich menu of suggestions made by experts and concerned citizens through the Track II dialogues in their official dialogue. The Indian government should take steps to address the grievances of its neighbours by extending the application of principles applied by it in respect of water-sharing issues and disputes among its states within the Union to its co-riparians of shared basins. South Asian governments need to share information and data on issues relevant to water.

South Asian countries should consider including water-related cooperation in the mandates of the regional inter-governmental organization such as the South Asia Association for Regional Cooperation (SAARC) and the South Asian Cooperative Environment Program (SACEP) as well as support regional non-governmental organizations and forums devoted to promoting cooperation on water-related issues. The establishment of a SAARC Centre on Integrated Water Resource Management in the framework of SAARC deserves to be considered.

Given "the hydrological dependence of the trans-boundary South Asian river basins on China due to the location of their headwaters in the South-Western region of the Tibetan Plateau", as well as China's plans to build multi-purpose dams, there is an urgent need for initiatives by China and the South Asian states on water issues.

Finally, South Asian states need to add the multifaceted impacts of climate change on their water resources in the frameworks for climate change-related cooperation, including the Thimpu Statement adopted by the SAARC Summit in 2010.

References

1. Ali, M.C. (1967). *The Emergence of Pakistan*, Colombia University Press, New York.
2. Bishti, M. (2008). *Revisiting the Moshe Agreement: Lessons for Indo-Nepal Water Diplomacy*, IDSA, New Delhi.
3. Biswas, A.K. (1992). Indus Treaty: The Negotiation Process, *Water International*, vol. 17.
4. Briscoe, J. (2010). Troubles Waters: Can a Bridge be Built over River Indus, Economic and Political Weekly, 17 December, vol. XLV, no. 50.
5. David Michel, D. and Sticklor, R. (2013). *Connecting the Drops: An Indus Basin Roadmap for Cross Border Water Research and Policy Coordination*, Stimson Center, Washington DC.
6. Deepak Gyawali, D. (2002). "Water Beyond the State: Resolving Conflicts with Institutional Pluralism", in P. Shahhadevan (ED.) "*Conflict and Peace Making in South Asia*", Lancers Publications, New Delhi.
7. Gulahati, N. (1973). *Indus Waters Treaty: An Exercise in International Mediation*, Applied Publications.
8. Hanasz, P. (2014). *Sharing Waters versus Sharing Rivers: The 1996 Ganges Treaty*, Global Water Forum.
9. India's River Linking Project (2003). *India's Water Diversion Plan: Its Impact on Bangladesh*, The New Nation, city.
10. Iyer, R. (2005). Indus Treaty: A Different View, *Economic and Political Weekly*, 15 July, vol. 40, no. 29.
11. Malhotra, P. (2010). *Water Issues Between Nepal, India and Bangladesh*, IPCS Special Report, city.
12. Pangare, P. (2004). *Hydro-Diplomacy: Sharing Water Across Borders*, IUCN, city.
13. Price, G. (2014). *Attitudes to Water in South Asia*, Chathham House, London.
14. Ramaswamy R. I. (2003). *Water: Perspectives, Issues, Concerns*, Sage Publications, New Delhi.
15. Ray, A. (2012). *India, Bangladesh: Water Disputes and Teesta River Diplomacy*, the International Security Network (ISN), Global Voices, city.
16. Ranjan, A. (2015). "Water Conflicts in South Asia: India's Transboundary River Water Conflicts with and Nepal", *BIISS Journal*, Vol. 36, Nov–Dec 2015.
17. Salman, M.A. and Uprety, K. (2003). *Conflict and Cooperation on South Asia's International Rivers: A Legal Perspective*, Washington DC.
18. Siwakoti, G. (2011). *Transboundary River Basins in South Asia: Options for Conflict Resolutions*, International Rivers: Kathmandu.

Chapter 4
Transboundary Waters: Why Does the People's Voice Lack Collective Echo?

Sudhirendar Sharma

One issue that literally crisscrosses divided Indian subcontinent is water: both in terms of the river systems that flow across national boundaries and the rain clouds that blow over it. Yet all across the subcontinent, water has become a source of nativist suspicion, distrust, and suppressed aggression. While India and Bangladesh have regularly squabbled over the waters of the Ganga and the Teesta, river treaties with other neighbors, Nepal and Pakistan, have also had their share of mutual distrust.

On the issue of river rights, India stands, as in so many matters of its international relations, in an in-between position that has the potential to trigger a plausible "water war".[1] On the one hand, it needs to be able to store and make use of river water that flows into Bangladesh; on the other it needs more transparency in its shared concerns regarding river waters with Nepal and Pakistan, and seeks to restrain the impact of China's actions on the Brahmaputra, as they may drastically affect flows in the Ganga–Brahmaputra basin.

Notwithstanding the multi-lateral regime of rivers in the sub-continent, the cross-boundary internal security threats have instead forced India to view the conflicting river issues from a bi-lateral lens. India's unilateral position with respect to setting up of a Ganga River Basin Authority to shift from the "current piecemeal efforts to a more integrated approach for ensuring quality and quantity of flow"[2] has only aggravated the prevailing suspicion towards its existing institutions engaged in river management.

Curiously, however, it will be fair to conclude that this region cannot escape the fact that it is part of the world devoid of any working institutions that can coordinate and integrate choices, and that can collectively confront predicaments faced by states and governments, and future challenges on account of climatic aberrations—

[1]While the popular "water-wars" thesis has largely been discarded, there are others who argue that violent conflict over water may still be possible. Researchers, however, have begun to replace "water-wars" with growing "conflicts of interest", which is quite common.
[2]Source: http://www.moef.nic.in/sites/default/files/ngrba/index.html.

I. Ahmed (ed.), *South Asian Rivers*, The Anthropocene: Politik—Economics—Society—Science 21,
https://doi.org/10.1007/978-3-319-67374-5_4

institutions that could sustain any degree of trust between neighboring states. In fact, in their absence, the subcontinent has remained vulnerable to the tyranny of geography.

4.1 The Omission of Commissions

No wonder, across borders there are joint river commissions to squabble over while the provisions of the inter-state dispute resolution mechanisms are enough to enrage states over their disputed share of river flows. Stressed by demand-supply inequities, the ecological entity of the river basin is held ransom by the warring factions. Be it Indus or Ganga flowing across international borders or inter-state rivers like Krishna or Cauvery, each of the major rivers has been at the receiving end of such political feuds.

Be it Islamabad, Kathmandu, or Dhaka or for that matter Delhi, Chennai, or Bangalore, in all these places people have become increasingly fanatic about what for long has been 'their' river, its water being shared across and along the basins. Undermining its collective existence, South Asia doesn't seem to have a common essence that stretches across the huge amorphous terrestrial and cultural space but has become an amalgam of many nation states, which, for their own diverse interests, may not wish to come together.

The tragic irony is that existing treaties and tribunals have neither resolved nor are likely to resolve the upstream-downstream predicament in the region. Unless political arrogance and public fanaticism are defused from the realm of shared rivers by replacing the hard wires with far more stable and pragmatic set of soft codes, the potential socio-ecological value of shared waters will remain grossly under-utilized.[3] The absence of political vision and institutional innovation stand as hurdles to drawing new contours on water-sharing.

Ironically, even the existence of a regional entity like SAARC has not been of help in taking up the issues of transboundary rivers because of its charter, which does not allow member countries to take up bilateral matters in its formal discussions. India's negative attitude towards multilateralism is pervasive despite the stress by independent reports suggesting "the need for efficacy of the bilateral cooperation because salience of water in India's relations with its neighbors will increase in the years.[4]"

---

[3]In his assessment, James Workman Stockholm Water Front, No. 1, (April 2011) argues that if transboundary cooperation remains confined to politicians, lethal conflicts will break out among herders, farmers, and fishermen who soon will have nothing left to trade or negotiate. Aren't India and Pakistan exporting virtual water in grain to Europe, while elite middlemen are skimming off the biggest margins by keeping hungry farmers at the bottom of the race by ensuring that water is kept artificially cheap till it is all gone?

[4]*Attitudes to Water in South Asia*. London: The Royal Institute of International Affairs, Chatham House, 2014.

One of these reports[5] has further argued that if South Asia remains in turmoil, cooperation will become difficult. The challenge before India in the coming years will be two-dimensional: to manage its water resources better and simultaneously to manage its riparian relations with its neighbors. Without doubt, it would be prudent for India to guard against a doctrinaire approach to the question of bilateralism versus regionalism. However, this can only be achieved if the 'fortress' mindset gets replaced by a collaborative approach towards management of resources.

Given the commonality of water resources, the commonality of their utilization, and the commonality of the emergent issues, there is clearly a strong case for meaningful interaction between scientific institutions and water management agencies across the region.[6] Unfortunately, however, while cooperation is often stymied by mutual suspicion at the governmental level, outcomes of civil society dialogues, also called Track II initiatives, have not gone too well with the governments either.

4.2 The Muted Multiple Voices

Since the early 1990s, a wide array of journalists' exchanges[7] and civil society initiatives among riparian states has been underway. In addition to generating enhanced understanding of situations on the ground, such initiatives produced credible outputs on the politics and knowledge in the form of video documentation, newspaper articles, joint declarations, and semi-academic research papers and manifestos. How far have such initiatives contributed to the regional policy environment, which continues to remain a matter of grave conjecture?

4.3 The Water Manifesto

One of the earliest of such initiatives was the Fellowship in South Asian Alternatives,[8] which brought out perhaps the first South Asian Manifesto on Water, Power, and People in 1997. The manifesto had sought to free "the young South

---

[5] *Water Security for India: The External Dynamics. New* Delhi: The Institute for Defence Studies and Analysis (IDSA), 2010.

[6] Michel, D. & Amit Pandya (Eds), *Troubled Waters: Climate Change, Hydropolitics, and Transboundary Resources.* Washington D.C.: The Henry L Stimson Center, 2009.

[7] The London-based Panos Institute had piloted journalists' travels across the length of some of the transboundary rivers, notably Ganga and Kosi, in Bangladesh, India, and Nepal in 1997. It created a fellowship of journalists who pursued the subject over the years, culminating into an authoritative multi-lingual publication *Disputes over the Ganga: A Look at Potential Water-Related Conflicts in South Asia* (2004). The present author has been part of both the initiatives.

[8] Ahmed, I, Ajaya Dixit, & Ashis Nandy. *Water, Power and People: A South Asian Manifesto on the Politics and Knowledge of Water.* Colombo: Regional Centre for Strategic Studies, 1997.

Asians from the stereotypes and clichés of the past," and to highlight that the construction-led water development in the region had failed to address the basic sources of the suffering of the people. Far from any transformation, water management is still pursued to enhance supply through technical interventions only.

Given the fact that the ownership of water is not vested in an individual or a group of individuals but in the state, developmentality and statism have continued to manifest themselves in technology-driven decision-making by the state often at the cost of painful 'dehumanization'. Despite a growing understanding that rivers need to be treated more holistically, contentions between states have remained focused on the volume of water flowing downstream, and rarely on how water is consumed within the river basin.

Through first decade of 2000, social activists under the aegis of the South Asian Solidarity on Rivers and People took reductionist engineering for river water control head-on, questioning the ecological veracity of several projects including the Tipaimukh Dam, and the proposal for Interlinking Himalayan and Peninsular Rivers. Unlike other similar initiatives, it engaged with activists and academics from member countries to debate and bring those issues into the public domain, which could not be admitted onto the official agenda.

Each of these Track 2 initiatives (as against governmental initiatives being termed Track 1) created space and scope for debating contentious issues, allowing multiple definitions and diverse perspectives on water governance to find place for different social proclivities and economic aspirations. Many of these initiatives[9] had a life of their own, borne out of social and intellectual compulsion of a group of people to bring a set of issues related to water governance to the attention of policy makers in the region.

4.4 Attitudes to Water in South Asia

Given the fact that discussion about transboundary water issues is often vociferous and somewhat antagonistic and with the earlier initiatives not making much impact on the policy arena, a project called "Attitudes to Water in South Asia"[10] was initiated to explore attitudes in South Asian countries to shared rivers in the region, to understand if "shared challenges" had made way for nationalist standpoints based on past injustice or perceived hostile notions.

Often to the contrary, unilateral decisions on shared water resources play out differently through the various systems of democracy across South Asia. Some 80% of the 500 water experts, policy-makers, and decision-makers who contributed to

[9]In the early 1990s, the Patna Initiative was a significant Track 2 congregation of academic and activist groups to engage in multiple track debates and discussions on the Himalaya–Ganga landscape. Many of the participants contributed to similar initiatives thereafter.

[10]*Attitudes to Water in South Asia*. London: The Royal Institute of International Affairs, Chatham House, 2014.

the study, hold a negative view on India's track record on transboundary water cooperation. The outlook on current trends, however, suggests that local grievances over water availability and quality will spread and intensify.

Populist measures like river-linking may hold short-term promise for one country but not without causing collateral damage to lower riparian countries, a fact that has been grossly overlooked. The Chatham House Report, a study conducted by the Royal Institute of International Affairs, places the zero-sum way in which water relations are viewed by diverse stakeholders throughout the region at the centre of the brewing discontent.

The study took a comprehensive look at a cross-section of opinions and perspectives on two of the most complex river systems in the sub-continent—the Ganga–Brahmaputra–Meghna basin that is home to some 630 million people and the Indus river basin in which an estimated 300 million people live. The respondents were unanimous that the region is facing acute water shortage because of rising demand, which will only escalate in the future.

If economic co-operation in the region is what the Indian Prime Minister has in mind then the interlinking of rivers is likely to water it down. Among other factors, economic growth demands investment for resource extraction and infrastructure development. Haven't economists long argued that water scarcity is probably one of the biggest risks for investors? It is in this light that river-linking is seen with serious apprehension in the region. Curiously, however, such apprehensions have hardly been officially acknowledged.

Of special significance is the fact that water issues play a key role in nurturing anti-Indian sentiment and in deepening hostile intentions among people of the bordering countries. Ironically, the idea of holding vital data and information on the hydrology of transboundary rivers from riparian countries on the pretext of national security may end up threatening security should anti-Indian sentiments be allowed to persist. The respondents felt that data sharing and dissemination could serve to ease transboundary water tensions.

For Bangladesh, the problem centres around upstream infrastructure development in the Ganges river basin; for Pakistan, water shortage in the Indus river basin is the cause for strained bilateral relationships; and for Nepal, unfulfilled hydro-power potential by India is the bone of bilateral contention. Unless India comes good on its water cooperation agenda, the foundation of regional cooperation will continue to remain weak at its core.

In spite of the shared river systems and interdependencies, governments in the region have only drawn bilateral water agreements whereas there is an urgent need for comprehensive regional understanding on sharing water sources in view of the rising demand on one hand and uncertainty of climate change in sustaining flows in glacial-fed rivers on the other. The onus for taking a lead in cementing bilateral relationships rests with India. However, some respondents held that unless long-standing internal disputes between provinces about water-sharing are resolved, addressing disputes with neighbouring countries would be presumptuous. The question that has remained unaddressed is: if domestic allocations and management are so difficult, how can countries expect to resolve disputes with neighbours?

Box 4.1: South Asia Water Initiative[11]

Having brokered the most durable Indus Water Treaty in the region, the World Bank initiated a multi-donor trust fund initiative in 2012, the South Asia Water Initiative (SAWI), to increase regional cooperation in the management of the major Himalayan river systems in South Asia, and to deliver sustainable, fair and inclusive development and climate resilience. The 5-year program aims to address poverty alleviation, gender and climate change issues in the context of river basin management, and support activities relating to transboundary water cooperation and investments in water resources management in all countries of the region. In addition to informing new or existing bilateral or multilateral governance processes, SAWI aims to secure new investments through bilateral or multi-lateral governance processes, and improve the quality of the planning processes through approaches that are more participatory and technically better informed.

With the Farakka Treaty between India and Bangladesh, signed in Dec 1996, on its last leg SAWI is working extensively on activities to improve the understanding of the hydrology, water resources and ecological health of the Ganges River, and to raise the awareness and acceptance of the importance of data and knowledge sharing to support evidence-based participatory water resources planning and development. This encompasses cooperation on a broader range of water-related benefits (including flood management, inland navigation and ecosystem services) in a basin context rather than simply focusing on lean season water sharing arrangements, SAWI is working actively through capacity building initiatives to support both technical and diplomatic capacity in both countries to enhance the quality and balance of water dialogue and negotiations. SAWI is demonstrating improved approaches to flood forecasting that may offer new opportunities for sub-regional water cooperation between India and Bangladesh as well as the upstream riparian countries (Bhutan, China and Nepal).

In the Indus Basin, SAWI is supporting the governments of Afghanistan and Pakistan as they explore options and opportunities for cooperative development of the water resources of the Chittral/Kunar and Kabul river basins, in particular for hydropower generation. SAWI has also undertaken technical studies to ensure planning processes are based on improved technical knowledge, and is working with both governments to explore appropriate joint governance arrangements and to build the professional and institutional capacity for joint governance.

[11]Supported by the UK, Australian, and Norwegian governments, the World Bank initiated South Asia Water Initiative (SAWI) in 2012 to support countries in improving and deepening transboundary dialogue, enhance the basin and water resources knowledge base, strengthen water institutions, and support investments that lead to sustainable, fair and inclusive development (http://www.worldbank.org/en/programs/sawi).

In terms of investments secured via bilateral or multilateral processes, there are emerging opportunities in different parts of the region that SAWI is positioning itself to actively support. In the eastern sub-region, discussions at multiple levels are exploring opportunities for investments to further develop the Ganges and Brahmaputra rivers for inland navigation in ways that protect or restore the ecological integrity of the rivers. Such developments would not only greatly support commerce and trade between countries, but would improve people-to-people connectivity and thus facilitate deeper regional integration. Similarly, discussions are progressing at multiple levels on investments to improve power grid connectivity in the eastern sub-region to distribute energy from new hydropower investments, with new power trading agreements that could benefit the four nations of the sub-region.

Improving water resources management is at the core of the initiative to deliver more economically efficient, socially equitable and environmentally sustainable use of water, and on mitigation of flood impacts, that together significantly contribute to poverty alleviation. Closely linked to poverty alleviation is gender, because the negative consequences of flooding, poor water management and increased water competition and pollution disproportionately affect women and children. Reducing the impacts on women in particular from flooding and poor water management can deliver significant health benefits, create opportunities for greater access to education for girls, and create greater opportunities for women to contribute economically through diversified employment opportunities.[12]

Source SAWI Annual Report 2015.

4.5 Water Voices

Recognizing the fact that rivers are a source of civilization, culture, religion, and the histories of societies settled for generations on their banks, another civil society initiative embarked upon capturing the collective but divergent concerns on the level of peoples' involvement in transboundary governance of as many as 2,000 households across 80 villages in Jammu and Kashmir (J&K), Uttar Pradesh, West Bengal and Bihar on the Indian side of the shared basins of the Sharda, Gandak, Kosi, Chenab, Kishan Ganga, Ganga, and Teesta rivers.

A part of similar studies being taken across shared river basins in Bangladesh, Nepal, and Pakistan, *Blues Beyond Boundaries: Transboundary Water Commons*

[12]Though the role of civil society initiatives in setting the SAWI agenda has not been officially acknowledged, careful reading through the objectives of SAWI reflects that several of those issues viz., poverty alleviation, basin integrity, risk forecasting, capacity building, joint governance and resource sharing have been factored in.

India Report[13] relies on painstaking surveys conducted over a period of two years to gather testimonies of river-dependent communities on the crises of both quantity and quality of flows in the rivers on account of various interventions and infringements, affecting local life and livelihoods. Based on this ground-breaking survey, the report provides details of the impact of river damming, flow diversion, and bank erosion on natural resources, water availability, and food security. The report shows that a large section of the surveyed households has little or no preparedness for facing human made disasters.

For example, Devi Raj of Pul Doda village of Doda district in Jammu and Kashmir laments the loss of productive land on account of the Baglihar Dam on the Chenab river; Maho Mandal of Nijtaraf village in West Bengal's Cooch Bihar district talks about reduced fish yields following the tempering of flows in the Tessta river and, Rabadi Devi of Pakariya village of Uttar Pradesh's Lakhimpur Kheri district offers a tearful testimony of the devastation caused by the barrage on the Sharda river. Rarely, if ever, have such voices of concern counted in transboundary negotiations.

Caught between the vortex of popular narratives and political overtures, a majority of respondents have blamed other countries for their water crises. Countries have signed bilateral water-sharing treaties on each of the rivers taken up in the study, but local people do not understand the technical details of these treaties on account of poor levels of literacy.[14] Though local people have adequate knowledge about the river and the impact caused by construction across its natural flow, local opinions are rarely taken into account in policy discourse on river water management, leaving the river-dependent communities high and dry.

Blues Beyond Boundaries adds to the growing literature from non-governmental organisations and citizen groups that aims to influence policies towards effective transboundary water governance, though these initiatives take place outside the official channels. But despite generating fresh insights, these studies have hardly been able to knock at the policy corridors. Were such studies designed to influence policies in the first place? The study admits that the search for solutions to manage contentious water-sharing issues is fraught with a set of political, emotive and divisive factors, and yet expects that that the independent initiative "contributes to formulation of policies in India, and the South Asia region".

Peoples' testimonies clearly reflect that river water-sharing agreements based on volumetric allocation have little appreciation of the needs of local communities living along the river. Unless there is a healthy relationship and cooperation among transboundary communities, striking a mutually acceptable deal for water-sharing may remain elusive. Despite the odds, the study succeeded in capturing voices from across the shared river basins for the first time. It highlighted the need to:

---

[13]*Blues Beyond Boundaries: Transboundary Water Commons India Report*. New Delhi: Action Aid Association, 2015.

[14]*Blues Beyond Boundaries*, op cit., 2015.

- *ensure* minimum flows to protect the right of the river;
- *share* the details of transboundary water treaties with local communities;
- *protect* women and children against human-induced disasters; and
- *uphold* and add value to the traditional livelihoods of river-dependent communities.

The study converted peoples' responses into 'numbers', to add credence to the findings and feed into the policy making process. Yet, such studies remain at the margins of policy planning process. Why do governments in the region remain blind to outcomes of such initiatives? How can civic engagement contribute to better water management in a region that is rapidly becoming water insecure?

It is evident that current water technologies adopted by governments such as big dams, diversions, and hydro projects have not met their stated objectives but have instead created discord in situations where harmony existed between communities across borders. Further these projects are witness to serious, long-term, and widespread negative ecological and livelihood impacts. Climate change has brought further challenges such as glacial melting, flash floods, landslides, droughts, forest fires, intermittent rainfall, increased sea levels and risk of salinity ingress. On top, the governments have not shown any use of basic values like equity, transparency, accountability, sustainability and participation of the people in intra and inter governmental processes.

From the foregoing narrative it is clear that the idealism of "soft options" in sharing transboundary waters is unlikely to work in the region whose political geography has cut into its cultural history. Torn apart by fanaticism and dominance over resources, the region may need to explore fresh perspectives and new approaches to address the water-sharing predicament. Given the current challenges that the region faces, we cannot confine water issues to nation states; only those approaches that bring people's perspectives to the centre stage can bring some semblance of order in equitable sharing of our rivers. The following questions and propositions are worth considering.

4.6 Can Scarcity Trigger Cooperation?

Many researchers contend that water scarcity is the basic motivation for conflict, and ultimately violence between states. Given that water is crucial for basic survival and is irreplaceable, states may even take up arms to defend access to shared river.[15] It has further been argued that resource poor regions are often incapable of creating institutions to manage conflicts, and consequently remain witness to exploitation at

[15]Falkenmark, M. *Water Scarcity Generates Environmental Stress and Potential Conflicts*. In *Water, Development, and the Environment*, edited by W. James and J. Niemczynowicz. Boca Raton: Lewis Publishers, 1992.

the hands of the dominant riparian. The relationship between scarcity and conflict is often drawn as linear.

However, co-relation between scarcity and conflict is somewhat convoluted and hence modified by other researchers,[16] who have argued that scarcity-conflict theory could work both ways: (1) conflict is least likely when the resources do not exist or exist in low quantities, and (2) inter-state conflict is most likely when the resource is moderately available. Though somewhat intriguing, the modified interpretation of conflict theory creates a paradoxical situation, that scarcity creates the potential for inter-state conflicts as well as opportunities for cooperation.

Examining some 271 bi-lateral treaties on 226 shared rivers signed between the years 1950 and 2002, Dinar[17] coined a relationship called *scarperation* (borne out of scarcity+cooperation), which proposes that the likelihood of cooperation increases with rising scarcity levels, but as scarcity increases beyond a certain level the incentive for cooperation diminishes. There is a catch, however, that to initiate cooperation, scarcity must be experienced by both parties. Yet, scarcity may not be an exclusive condition for cooperation.

There are additional variables that explain *scarperation* among two riparian states sharing a river. Governance levels in the basin are also significant in explaining levels of cooperation. However, the strength of cooperation is often determined by how many 'issues' are addressed in the treaty. In the past, a majority of treaties were single-issue ones and hence largely ineffective. In recent years, however, multiple issues namely, water allocation, hydropower production, pollution control and flood protection have been addressed in treaties, making cooperation between riparian countries last longer. Research on the subject is in the early stages, but creating theoretical models of scarcity based on projected water demand across the basin can be used to engage the governments in re-writing existing treaties to factor in issues that may lead to scarcity.

4.7 Can Trade Reduce Conflict?

The literature is quite mixed in its assessment of links between trade, conflict and cooperation, though the extent of trade between states influences the scope of their diplomatic relations. It has been generally claimed[18] that increased trade between states reduces incidents of militarized conflict between them, and promotes peace. The fear of losing gains from trade deters conflict. However, there is another school

[16]Giordano, M., M. Giordano, and A. Wolf. *International Resource Conflict and Mitigation.* Journal of Peace Research (2005), 42:47–65.

[17]Dinar, S. *Scarperation: A theory of Scarcity and Cooperation Over Transboundary Rivers.* Department of International Relations and Geography, Florida International University, Working Paper, Miami, 2006.

[18]Russett, B., and J. Oneal. *Triangulating Peace: Democracy, Interdependence, and International Organizations.* New York: W.W. Norton & Company, 2001.

of thought that argues that high levels of inter-state trade increase the chances of friction among the countries, leading to possible conflict.

In the context of both the trade-conflict and trade-peace corollaries, it can be deduced that inter-state trade indicates not only a history of cooperation between countries but also the possibility of achieving negotiated settlements. Trade, it seems, also acts as a contract enforcing mechanism. Even if there is a likelihood of disputes between states, it has been argued that trade interests provide an opportunity for states to resolve disputes at a lower level of interstate conflict. In a nutshell, the coercive potential of trade is known to reduce conflict.

Barring exceptions, trade relations have a positive impact on the effectiveness of environmental treaties. The stronger the relations between states, the better the chances for treaties to stand the test of times. However, trade relying on the import or export of virtual water via traded goods may have varying impact. Subject to how the states perceive trade in such goods, virtual water trade can have deleterious effect on the lives of farmers and fishermen who stand to lose livelihoods on account of the terms of trade which invariably benefits the middle-men. At the other end, there are possibilities that trading countries may reduce the pressure on their scarce water resources by switching to goods that do not relate to trade in virtual water. However, trade in the context of transboundary cooperation should be considered in future research.

4.8 Could Cultural Threads Promote Interwoven Accords?

Drawing new contours is as much about shaping intellectual frameworks as about producing maps that can define new conceptual orders. Such contours will cut across political boundaries, creating distinct cultural spaces unified through a common heritage of religion, language, and social practices. Drawing some of the most powerful images of rivers, Rabindranath Tagore had argued that cultures changed a bit but continued to persist along the flow of the river.

Multiple cultural maps of South Asia, superimposed on the river basins, can bind communities with common cultural threads. While a basin map capturing the common Hindi chord along the trans-boundary rivers could infuse new hydro-solidarity amongst riparian communities in Nepal and India, a sub-basin map contoured around the Bangla speaking populace could easily build a better appreciation for upstream-downstream concerns between riverine communities of India and Bangladesh.

Much work needs to be done in drawing such maps along other cultural pockets. No longer can trans-boundary "environment flows" remain the political horse that pulls the economic cart. Instead, it ought to be bubbled up from below by people who risk everything because they have nothing left to lose. The opportunity of exploring pockets of cultural homogeneity seem promising, provided there is political will to depart from past failures.

It is undoubtedly clear that only by liberating common cultural strands from the bindings of the nation-state can a new order be imagined where free flowing rivers will be without guards or fences with watchtowers. Given the enormity of challenges confronting the region, rebuilding water security for teeming millions would need imaginative insights to shape intellectual frameworks that will blur martial nationalism of the kind that is currently all-pervasive.

4.9 Conclusion

In the context of the tyranny of geography, the search for a "river of cooperation" has to be found beyond the realm of politics of vested interests. From the Mekong and Columbia to the Indus, Volta and Limpopo, in the words of James Workman,[19] big changes in the management of transboundary rivers always have bubbled up from below from illiterate fishermen, floodplain farmers, and displaced families—people who risked everything because they had nothing left to lose.

Taking the argument further, Workman argues that much of research on transboundary river management is conducted for technocrats and offers an authoritative, cohesive, and integrated analysis of real problems based on previous experiences and potential solutions. However, there is clearly scope for expanding the circle of decision-making by including all basin stakeholders. There is a need for drawing a balance between abstract doctrine and basic instincts, such that interests in transboundary cooperation reach beyond the current elite.

[19]*In the Heart of Dryness* (http://www.heartofdryness.com) James G. Workman presents the timeless struggle over water in the Kalahari Desert in perspective. While indigenous Bushmen used voluntary survival strategies to fight drought, scarcity and climate change, the Botswana's government sought to enforce regulatory rule instead. Their rivalry foreshadows our world, where two in three thirsty humans will soon endure shortages, and contribute to a global fight for water as a human right.

Chapter 5
Trans-Boundary Water Governance in South Asia: The Beginning of a New Journey

Ajaya Dixit

In February 2016, Government of Nepal's Ministry of Population and Environment submitted the country's Intended Nationally Determined Contribution (INDC) to the United Nations Framework Convention on Climate Change (UNFCCC). Early on in the report, the INDC mentions, "In Nepal's Himalaya, total estimated ice reserve between 1977 and 2010 decreased by 29% (129 km^3). The number of glacier lakes increased by 11% and glaciers recede on an average by 38 km^2 per year." The INDC makes no reference to the study that made this claim, but subsequent literature search revealed that the Kathmandu-based International Centre for Integrated Mountain Development had completed the study. Its report *Glacier Status in Nepal and Decadal Change from 1980 to 2010* notes,[1]

> In 2010, a total of 3,808 glaciers were identified with a total area of 3,902 km^2 and estimated ice reserves of 312 km^3...The total glacier area decreased by 24% between 1977 and 2010 [33 years], and the estimated ice reserves by 29% (129 km^3).

If 1977 is taken as the reference year, the 129 km^3 decrease in ice reserves means that in the 33 years till 2010, the tributaries of the Ganga River were deprived of an average base flow of 123 m^3/s. If we assume that the recession continued at the same rate between 2010 and 2015, an additional 15.7 km^3 of ice reserves would have been lost, leaving Nepal with a balance of just 113.4 km^3. What will be scenario if the ice reserves continue be lost at this rate beyond 2015? A simple calculation would show that the reserve might be gone permanently in a few decades from today. This scenario should make everyone sit up because of its serious consequences. Any loss of Nepal's ice reserves would be catastrophic for the inhabitants of Ganga River Basin (GRB), home to some of the world's poorest and most vulnerable people. The rivers of Nepal generate 70% of the dry season flow of the Ganga River at Farakka, much of which is derived from snowmelt.

But discussion on such a scenario should remind us of the 2007 controversy on snow and glacier melt in the Himalaya Region. That year's Intergovernmental

---

[1]Bajracharya, S.R.; Maharjan, S.B.; Shrestha, F.; Bajracharya, O.R.; Baidya, S. (2014) *Glacier status in Nepal and decadal change from 1980 to 2010 based on land sat data*. Kathmandu: ICIMOD.

I. Ahmed (ed.), *South Asian Rivers*, The Anthropocene: Politik—Economics—Society—Science 21,
https://doi.org/10.1007/978-3-319-67374-5_5

Panel on Climate Change (IPCC) in its report had mentioned that Himalayan glaciers could melt away by 2035. Several senior scientists had opined that the claim was unrealistic and that the Himalayan glaciers could not melt in a few decades. Subsequently, the IPCC admitted its mistake and in a statement said the findings were based on "poorly substantiated estimates of rate of recession and date for the disappearance of Himalayan glaciers. In drafting the paragraph in question, the clear and well-established standards of evidence, required by the IPCC procedures, were not applied properly."

The alterations in behaviour of snow systems due to climate change, recession of ice reserves, and its consequences are complex processes. Research about the role climate change plays in snow systems, glacier dynamics, and related processes is incomplete. They depend on the dynamics of snow, glacier mass, precipitation, and other change processes as well as assumptions used for analysis. It would be therefore necessary to exercise caution in using such assessments for policy even though, climate warming will affect snow systems, increase the frequency of extreme weather events, heat and cold waves, and hazards such as floods and droughts.

Rivers in South Asia, particularly those in the GRB, face many challenges other than just climate change. The cover page of the book *Dirty, Sacred Rivers: Confronting South Asia's Water Crisis* by environmental journalist Cheryl Colopy makes a pertinent observation:

> Rivers, traditionally revered by the people of the Indian-Sub continent have in recent decades deteriorated dramatically due to economic progress and gross mismanagement. Dams and ill-advised embankments strangle the Ganges and its sacred tributaries. Rivers have become sewerage channels for burgeoning population.[2]

In GBR the degradation of natural sources is concurrent with three phenomena with serious implications for water bodies, livelihood, and wellbeing. The three factors are growing urbanization, dependency on food imports, and socio-economic structure. In the coming decades, a substantial portion of the GRB's population is more likely to live in urban than in rural areas, and social disparities and inequalities in income, opportunities, health, and education are likely to be substantial.

Urbanization is likely to transform landscapes as the current unsustainable practices of resource extraction that lead to environmental degradation and pollution are perpetuated. In addition, urbanization, by changing water-use patterns and demands, will have significant local implications. Pollution of water bodies with untreated wastewater and competition with other uses, such as irrigation and ecosystem services, will continue to increase. Urban areas are likely to become more dependent on services produced by distant than local ecosystems.[3] For

---

[2]Colopy, C. (2012). *Dirty, scared rivers: Confronting South Asia's water crisis*. USA: Oxford University Press.

[3]DST (2008). *Re-imagining the rural-urban continuum: Understanding the role ecosystem services play in the livelihoods of the poor in Desakota regions undergoing rapid change*. Research Gap Analysis prepared by the Desakota Study Team (DST) for the Ecosystem Services for Poverty

example, the eastern part of the GRB is a food-deficit area which, in a sign of growing dependency, must import food from other regions.[4] This shift toward dependency will entail new sources of vulnerability, as India's 2016 "unofficial blockade" prevented Nepal from accessing sufficient supplies of basic commodities, including food, fuel for vehicles, cooking gas, and medicines for which it had increasingly been dependent on India. In addition to the threats of urbanization and dependency, prevailing challenges emanate from a combination of low landholdings, food insecurity, poverty, unreliable energy supply, a lack of non-agricultural livelihoods, frequent floods, weak markets and weak local institutions.[5]

The risk of climate change, particularly on the regions' poor and vulnerable, requires a new approach and beginning to water resource use and management. If there is no new start, water systems will continue to be degraded and abused, exacerbating existing problems and creating new risks. Current efforts at developing and harnessing both in country and transboundary water must take cognizance of emerging local contexts and link proposed interventions to these realities.

5.1 Resource Base

The Ganga River, whose major tributaries originate in Nepal and Tibet, flows through the Indian states of Uttar Pradesh, Bihar, and West Bengal before entering Bangladesh, where it joins the Brahmaputra and the Meghna rivers and eventually empties into the Bay of Bengal. The Ganga drains an area of about 1.1 million km^2 and every year dumps a colossal amount of fresh water into the sea after passing through the world's largest mangrove ecosystem, the Sundarbans, on the river delta.

Thus any effort to harness the rivers of GRB collectively by the countries must recognize the above mentioned constraints if it is to generate positive outcomes. Low-income households must have greater access to water for livelihoods and productivity through access to irrigation. Access to water is a necessary but insufficient condition for securing their social and economic wellbeing. The ongoing efforts at managing and governing transboundary water have sought solutions in the manipulation of rivers to achieve benefits but without success. Rather than seeking solutions within waterscape, efforts must be made in

(Footnote 3 continued)

Alleviation (ESPA) Programme of Natural Environment Research Council (NERC), Department for International Development (DFID) and Economic and Social Research Council (ESRC) of the United Kingdom. Kathmandu: Institute for Social and Environmental Transition-Nepal (ISET-N).

[4]Sharma, B. and Vyas, S. (2013). Moving from water problems to water solution: An overview of the constraints and opportunities in the Eastern Gangetic plains. *Water problems to water solution.* Moving from water problems to water solutions: Research needs assessment for the Eastern Gangetic Plains. New Delhi: International Workshop.

[5]Ibid.

broader policy space that recognizes the nexus among water-energy-agriculture-livelihood.[6]

About 79% of the GRB area lies in India, 13% in Nepal, 4% in Tibet, and 4% in Bangladesh.[7] The area of the Ganga basin in India constitutes about 26% of the country's total land area.[8] Though Nepal and Bangladesh occupy small fractions of the basin, all of Nepal and about one quarter of Bangladesh lie within the Ganga basin. In India, the Ganga and its tributaries flows through or along the states of Uttarakhanda, Uttar Pradesh, New Delhi, Madhya Pradesh, Bihar, Jharkhand, and West Bengal. Because the Indian Constitution makes the use and management of water and energy state, rather than national, matters, the use, management, and governance of the river acquires an additional layer of jurisdictional complication.

The Ganga has two main types of tributaries, those originating in the Himalayan Range in the north and those originating in the Vindhya Range in the south. The tributaries in the north can be further sub-divided again into two types: those that originate above the snowline of the Himalayas and those below the snowline and fed by rain and groundwater. The major snow-fed rivers of Nepal and India are the Yamuna, Ganga, Ramganga, Mahakali, Karnali, Gandaki and Koshi. The rain-fed rivers of the north include the Kankai, Kamala, Bagmati, West Rapti, and Tinau in Nepal, and the Gomti and Buri Gandak in India.

River flow is at its peak during the four months of the monsoon (June–September), when the region receives about 80% of its annual rainfall. The rainfall leads to a corresponding peaking in river flows from June to September. On average, the basin receives around 1,200 billion m^3 of precipitation of which about 500 billion m^3 flows out as surface runoff. The rest recharges groundwater aquifers or returns to the atmosphere through evapo-transpiration. These aggregated values do not show the immense spatial and temporal differences that characterize the sub-basins of the GBR. While glaciers and snow contribute only a small fraction of the total flow (just 10% of river flow), they are crucial in that they store the water that contributes to low flow. The rivers flowing from Nepal accounts 47% of the average flow of the Ganga at Farakka which increases to almost 70% in the lean seasons.[9]

The Ganga tributaries carry a large amount of naturally eroded sediment from the inland landscape and transfer it to downstream areas and eventually to the sea. High sedimentation rates are a natural consequence of the hydrology, geology, and topography of the GRB and are responsible for the continually changing river

[6]Shah, T. (2013). Research to lead development in Eastern Gangetic Plains: Some lessons from a decade of IWMI-Tata water policy. *Water problems to water solutions*. Moving from water problems to water solutions: Research needs assessment for the Eastern Gangetic Plains. New Delhi: International Workshop.

[7]Pun S. B. (2004) Overview Conflict over the Ganga? In Disputes over The Ganga, A Look at Potential Water Related conflicts in South Asia ed Subba B and Pradhan K. Panos South Asia, Kathmandu.

[8]Ibid.

[9]Ibid.

morphology. The bulk of the sediment is derived from point sources[10] such as the landslides and mass movements that occur randomly over the landscape. Such point sources are common on forest and agricultural land as well as on barren slopes. The Himalayan mountain range has an early infinite soil mass available for transport by rivers and river transport capacities are particularly high in the upper reaches where gradients are greatest. The soil mass that enters a river determines the geo-morphological behaviour of its flows and harmful impacts it has on the viability of flood control structures, embankments, irrigation systems, hydropower plants, and storage reservoirs.

Glacial Lake Outburst Floods (GLOFs) and landslide dam outburst floods add to the high sediment load. The Nepal Himalaya range alone has 2,323 glacial lakes. In just 76 years, between 1935 and 2011, there have been 24 GLOFs, about one every three years. GLOFs devastate local communities and significantly increase the sediment concentration of downstream rivers. In the Mid-Hills, a landslide which slides into a river may create a temporary "landslide dam," which, when it breaches, can result in an extreme flood, causing much local devastation thereby becoming a point source for a large sediment load. While mostly such "landslide-dam floods" occur during the monsoon, it is not uncommon for them to occur in other seasons as well. Their breach may lead to localized devastation.

The surge of sediment associated with GLOFs and landslide-dam floods has serious implications for hydraulic structures, as deposits lower the capacity of canals and water storage systems. When sediment-laden water flows through a turbine, its blades are damaged by contact with particles and the power output of hydropower plants may decline, thereby necessitating a temporary shutdown to facilitate repairs. Any such closure affecting the supply of energy will have an economic impact. Efficient flushing mechanisms must be designed, operated, and maintained if the likelihood of cessation of services is to be minimized.

The high concentration of sediment in rivers may also exacerbate bank erosion and result in the accumulation of sandy deposits on cropland. Such deposits result in long-term devastation as the land grows increasingly infertile and ultimately turns barren and must be taken out of production. Unfortunately, major stakeholders who participate in the discussions about water in GBR do not recognize sedimentation as a major risk. This is despite the fact that the extreme rainfall events and shifts in precipitation patterns and stream flow characteristics associated with climate change can increase sediment mobilization and yield.

While extreme floods are often catastrophic, annual flooding is highly beneficial as it deposits the fine silt that replenishes soil fertility, Furthermore there would be no highly fertile deltas, doabs, or flood plains and therefore, no agriculture if the rivers were not to transfer and deposit sediment. This annual cycle of fertility replenishment is one of the sustenance bases of the more than 655 million people who live in the GRB, spread out at an average population density of 551 people per km^2.

---

[10]Carson, B.(1986). *Erosion and sedimentation processes in the Himalayas*. Occasional paper No. 1. Kathmandu: International Centre for Integrated Mountain Development (ICIMOD).

Despite this agricultural boon, poverty is widespread. The example of India is a case in point. According to World Bank about 21.3% of the population has a poverty head count ratio at US $ 1.90. It is clear that majority of the population live in GBR region and the percentage would be higher if a criterion of multidimensional poverty were used.

5.2 Changing Political Geography and Transboundary Trajectories

As the political geography of a region changes over time, the claims on the benefits that can be derived from a transboundary river also change. In 1978, the United Nations determined that 214 basins in the world had a transboundary character, meaning two or more countries shared them. Today, that number has swollen to over 260 river basins and nearly 270 groundwater aquifers.[11,12] The GRB, like other basins, has also seen changes in its underlying tectonics, and geographic, and riparian relationships over the years.

Hundreds of millions of years ago, South Asia was part of the massive super-continent Pangaea. At the end of the Precambrian eon, a landmass that corresponds to present-day South Asia, Australia, Antarctica, and South America separated from Pangaea to form Gondawana. Between Gondawana and Eurasia, which included the present landmasses of North America and Asia, was the Tethys Sea. In the late Cretaceous era the landmasses comprising Gondawana began to separate and move in different directions. The South Asian landmass moved northwards and eventually collided with Eurasia. The collision caused the Himalayan, Mahabharata and Chure (Siwalik) ranges to rise along the northern section of the landmass. North of the Himalaya, the Tibetan Plateau was formed. In this tectonic transition, the Tethys Sea disappeared. This geological sequence, which has always been of interest to geologists and seismologists, became an issue of wider concern in the aftermath of the 2015 Gorkha Earthquake that hit central Nepal. The Indian tectonic plate, as the South Asian landmass is called, continues to move north and push against the larger Eurasian plate, making the Himalaya tectonically very active and increasing the possibility of major earthquakes.

Subsequent to these topographical changes, the monsoon dynamics also began in South Asia. In the beginning of April, when the temperature over the Bay of Bengal and the Arabian Sea is low, the Asiatic mainland heats faster than the water, producing a difference in atmospheric pressure that forces the warm, moisture-laden air mass over the sea to flow towards the South Asian landmass. The orographic

---

[11]Wolf, A. T., Natharius, J. Danielson, J., Ward, B. and Pender, J. (1999). International river basins of the world. *Int. J. Water Resource. Dev.*, 15(4), 387–427.

[12]UNESCO (2009). *Atlas of transboundary aquifers: Global maps, regional cooperation, and local inventories*. Paris, France: UNESCO International Hydrological Programme.

effects of the mountain ranges then cause the water vapor to cool and condense, causing rain, snow, or other form of precipitation to fall on the land, where it feeds various sub-processes of the hydrological cycle before flowing into the sea. The channelized flow of water from the Himalayan range led to the formation of the Ganga, Indus and Brahmaputra rivers and their tributaries. These rivers, like all rivers, serve as natural conduits that transfer water from inland areas to the sea in a crucial sub-process of the regional hydrological system. The flood plains of the Ganga and Indus rivers south of the Himalaya and lower mountain ranges were once grasslands inhabited by the prehistoric ancestors of elephants, tigers, and other animals. Eons later, these plains became part of various kingdoms and empires, including the Harappa and Mohenjodaro civilizations, which flourished several centuries before the Christian Era; the Sunga, which saw its heyday in the first century BCE; and the later Pali, Mughal, and British empires, before the countries of South Asia became independent. Today, these plains are home to millions of Nepalis, Indians, Pakistanis, and Bangladeshis.

The rivers in GRB support agriculture, fishery, and other water-based livelihoods, and sustain millions through the ecosystem services they provide. Commerce flourished along the banks of its many rivers as inland navigation helped promote trade. The importance of the rivers for livelihood and economic development gradually became intertwined with religious, cultural and social norms that recognized and celebrated their value. Soon, rivers had acquired a sacred quality and temples and shrines were built on their banks and at their confluences. Rivers, after inspiring poets, saints, thinkers, writers and artists, became part of folklore, classic and modern literature, and rituals. Many spiritually-minded people claim to understand the nature of reality itself by meditating on riverbanks. Riverbanks are also seen by Hindus as the ideal location for cremation: they believe that mingling a person's last mortal remains with a river's flow will secure the salvation of his or her soul. Every religion practiced in South Asia—and there are many—reveres water in its different forms.

With the colonization of the Sub-Continent, the rivers of the GRB also acquired a political dimension, and their water began to be used to meet political objectives. The process of politicizing rivers involved conducting negotiations and signing agreements among various users, including, in modern times, neighbouring countries. In South Asian history, however, the political objective became dominant once the British East India Company and later the British state began to develop irrigation systems. The aim was to generate revenue: if irrigation canals could provide sufficient water to increase agricultural production, more tax could be collected. The colonial state had two other aims as well: maintaining administrative control within the irrigated area and providing supplemental irrigation in case of a drought or extended period without rainfall. The Upper Ganga Canal (UGC) in the north Indian plain, which was completed in 1854, was one of the first major irrigation canals the East India Company built. Later, it began building irrigation systems that tapped the tributaries of the Indus River in order to irrigate the Indus plains. In 1869, the government of the United Province (now Uttar Pradesh) proposed to use the Sharada River (the Mahakali in Nepal) to irrigate the Awadh plains by constructing a barrage. The barrage provided with metal gates would regulate river flow and divert water into canals for irrigation.

About a century prior to these efforts, the political geography in the Himalaya region had begun to change as Prithvi Narayan Shah in 1768 unified the *Baise* (22) and *Chaubise* (24) principalities west of present-day Kathmandu Valley into Nepal as we know it today. The Shah dynasty began to expand its boundaries and ultimately collided with the British East India Company, which was also expanding its power in the Ganga and Indus plains by occupying the void left by the disintegrating Mogul Empire. The British East India Company and Nepal fought many battles. In the decisive battle at Nalapani, Nepal lost, thereby paving the way for the signing of the Sugauli Treaty in 1816. This treaty established peace between British India and Nepal and defined Nepal's boundaries till the 1850s. The relationship between the neighbours reached a new dimension during the 1857 Awadh Sepoy Rebellion (known in India as the First War of Independence), when Nepali Prime Minister Jung Bahadur dispatched the troops that successfully controlled the rebellion. In return for Nepal's help, the British government returned the districts of Banke, Bardiya, Kailali, and Kanchanpur to Nepal. The Sharada (Mahakali) River drains areas in West Nepal and the hills of Kumaon before debouching onto the plains at Brahmadevmandi, Kanchanpur District, which borders India's Uttar Pradesh (then United Province).

The government of United Province identified the eastern bank of the Mahakali River in Kanchanpur District as a suitable site for building a Barrage. To build the Sharada Barrage the government of United Province sought to acquire about 4,000 acres of Nepali forestland. It began negotiating with the government of Prime Minister Chandra Sumsher in 1910, but it was not until ten years later, in 1920, that an agreement was finally signed between the two governments. This agreement, which was the first treaty ever negotiated in South Asia regarding a transboundary river, allocated to Nepal 150, 450, or up to 1,000 cusecs of water depending on the month—if it was available. The agreement was revised twice, in 1991 and 1992, and then subsumed under the 1996 Mahakali Treaty.

In 1947, almost twenty years after the Sharada Canal system was commissioned, British colonial rule in South Asia ended. India and Pakistan, once both part of the British Empire, became two independent countries. Their independence changed the geo-politics one more time and, with it, the nature of engagement in transboundary rivers. The departure of the British also spurred political changes in Nepal: the hereditary Rana dynasty was abolished and a multiparty democratic polity established. The next treaty on a transboundary river was not signed until 1954, when Nepal and India agreed to share the waters of the Koshi River. The Koshi Treaty, which was revised in 1966 and 1978, included provisions regarding flood control, irrigation, and hydropower benefits for both countries but was perceived by Nepalis as being unequal, with India getting more than its fair share of benefits. In 1959, Nepal's democratically elected government signed a treaty on another transboundary river, the Gandak, and in 1960, India and Pakistan signed one on the Indus River. The Indus Treaty is presented as a success because it has survived wars between the neighbours though it is argued that it partitioned the Indus and its tributaries between India and Pakistan.

With the march of time, yet another geo-political change was to take place in South Asia and a new transboundary context emerged. In 1971 the erstwhile East Pakistan became the independent country of Bangladesh and passed on its dispute with India about the sharing of the waters of the Ganga at Farakka. In 1962 India had begun constructing a barrage at Farakka to divert a portion of the Ganga's flow into the Hoogly River in order to maintain its navigability. It signed a treaty with Bangladesh about sharing the flow at Farakka in 1977 and again in 1996, the same year that India and Nepal signed the Mahakali Treaty, which incorporated the 1991 and 1992 memoranda of understanding on the Tanakpur and Sharada barrages. Since then, the political geography within India has undergone several changes: Uttarakhanda, Jharkhanda and Chhatisgarh have all emerged as new states within the GRB. We can expect more changes within Nepal, too, as the country's new constitution stipulates that the country will be a federal democratic republic. Though the demarcation of Nepal's federal provinces has not yet been settled, it is clear that the riparian context will change.

Thus far, however, the treaties on transboundary rivers have had only a limited impact on socio-economic development: indicators like GDP, poverty rates, and access to services in the countries comprising the GRB, lag far behind those in many other regions of the world. As climate change proceeds, the GRB region is likely to be stressed by a range of impacts, all of which will exacerbate the challenges to overcoming poverty that nations already face. The complex climate dynamics range from glacial processes to flash flooding and drought brought on by excess and deficient monsoon rains respectively. Climate change impacts including declining snow cover and glacial retreat, more variable and more intense extreme rainfall events, and rises in sea level, temperature and heat content are likely to exacerbate the vulnerability of the inhabitants of the GRB. Climate change dynamics will compound the already high levels of exposure to the floods, cyclones, snowstorms, droughts, landsides, and other hazards that already prevail across virtually all parts of the GRB.

The annual recurrence of floods, droughts, and storms results in shortages of potable water, food insecurity, loss of livelihoods, high rates of mortality and morbidity, and disease outbreaks, as well as invasions by exotic plant and animal species, which wreak havoc in ecosystems and reduce biodiversity. Impacts on poor rural populations that depend primarily on little diversified agricultural systems for their livelihoods are particularly adverse. The limitations to resilience inherent in climate-vulnerable livelihoods, particularly agriculture, intertwine with social and economic realities like exclusion and gender and ethnic differentials to push large sections of the population further into marginalization, poverty, and debt.

5.3 Natural System and Diverse Water Uses

The use, development, and management of water in the GRB are embedded within the complex dynamics described above. The natural dynamics of water is the outcome of interactions between the Himalayan range and the South Asian Monsoon (SAM) in the summer and the westerly winds in the winter, the interactions which result in conditions of too much and too little water respectively, as normal annual phenomena. The GRB region as a whole receives most of its water during the summer monsoon. The basin's outlet, Bangladesh, receives higher rainfall and its amount decreases steadily as one moves inland though certain inland regions, like the Middle Mountain region of Nepal, are exceptions. In central Nepal, south of the Annapurna and Dhaulagiri mountains, annual rainfall is 5,000 mm. In contrast, the average annual rainfall over Nepal is just 1,800 mm. Precipitation in the form of snowfall maintains ice and glaciers, while rainfall recharges springs and groundwater, augments stream flows, and replenishes lakes and wetlands. These water bodies support local-level livelihoods and economies.

Upstream diversion, land-use changes, and climate change-induced impacts have altered the sub-processes of the regional hydrological system. In particular, stream flow variability—the difference between maximum and minimum flows—is changing. Seasonal changes as well as the occurrence of extreme events increase the risk of flooding and drought and have negative implications for the constancy of hydropower generation and the reliability of irrigation and domestic water supply systems. More serious risks include declines in water quality, particularly when low flows increase pollution loads along urban settlements. Both low and high flows have implications for storage concerns: low flow does not fill existing storage facilities while high flow (floods) results in higher rates of sediment mobilization and thereby reduces storage volumes. The implications of these changes for ecosystem services and the people who depend on them can be serious but the nature of these changes has yet to be examined critically.

Spring flows are important because the majority of people living in Mid-Hills region of Nepal and India rely on them for meeting their everyday needs. People report that in many places spring flows are declining and there is not enough water to meet even drinking needs. As the supply declines, the community systems built to tap springs function sub-optimally, forcing users, particularly women, to travel to distant and low-quality sources. The decline in spring flows can be attributed to changes in the intensity and frequency of rainfall, the snow-rainfall mix, long dry periods, and non-climatic factors such as road construction and changes in land use.[13] Recent research shows that the condition of aquifers and water-use patterns are related, but the impact of climatic factors on this relationship has not yet been

---

[13]Dixit, A, Gyawali, D., & Pandey, U. (2012). Improved access to safe drinking water, sanitation and health: Foundations for adapting to climate change impacts in Nepal. In V. I. Grover (Ed.), *Impact of climate change on water and health* (pp. 200–247). USA: CRC Press.

examined systematically. So far, research designed to unpack this interdependence has been minimal.

Catchment rainfall, springs, stream headwaters, and tributaries play significant roles in the recharge of groundwater aquifers. The porous formation of the Bhabar zone south of the Chure hills recharges aquifers in Nepal's Tarai. These aquifers are connected to those in the plains of northern Uttar Pradesh and northern Bihar. The large amounts of water available in these underground reservoirs means that the GRB is relatively unaffected by short-term rainfall variability: this supply provides buffering capabilities that ensure a steady supply of water for both irrigation and domestic use, even during dry periods. In fact, in some parts of South Asia, the use of groundwater has helped stabilize agriculture and increase household incomes.[14] Current estimates put the total number of electricity- and diesel-powered groundwater pumps in the GRB at 25 million.[15] This significant shift toward stability and even prosperity was accomplished by virtue of the fact that the governments of countries in the GRB provided farmers with subsidies to pump groundwater. This unfortunately has had two significant externalities. The first is the high fiscal hemorrhage the governments have experienced in trying to meet the cost of providing subsidies. The second is that excessive pumping of the groundwater has lowered water tables and degraded water quality. In many pockets of Nepal, in the Indian states of Bihar and West Bengal, and in Bangladesh, arsenic concentrations in groundwater supplies exceed WHO-prescribed safe levels and seriously threaten the health of local populations. This lowering of groundwater tables and contamination of groundwater cancel out the poverty reduction benefits associated with having access to a reliable supply of water and adversely affect the integrity of local ecosystems and the quality of their services. Identifying a direct causal relationship among groundwater use, climate and water management, however, is not easy because the natures of both aquifers and users are diverse. How the water-energy-food nexus will play out in the context of the increasing vulnerability of groundwater resources to over-pumping as well as other drivers is not clear.

5.4 Reflections on Policy Processes

Why poverty is pervasive in the GRB despite the abundance of surface and groundwater and other natural resources has been a subject of many enquiries by academics, intellectuals, and development professionals. Their analyses offer insights into how the gap could be bridged. While approaches to groundwater use and management have a broad interdisciplinary perspective, those to surface water

---

[14]Moench, M. (2003). Groundwater and poverty: Exploring the connections, *intensive use of groundwater: Challenges and opportunities*, Llamas, R., Custodio, E., Lisse, A. and Balkema, A. (Eds.), 441–156.

[15]Ibid.

use and management have been conventional: they have broadly perpetuated strategies inherited from British colonial administrators.

When India and Pakistan began their independent journey to water development they used the same approaches and methods that the colonial state had used. They favoured these methods because they had already been tested and seemed less risky. The approach the two governments followed considered issues such ecosystem, gender, and social diversity peripheral. With American President Harry Truman's 1949 declaration of his four-point agenda for the development of the Third World came the beginning of an era of generous development assistance for third world governments which rested on the belief that technology was a panacea for achieving the social and economic transformation of the least developed and developing countries. Thus, the conventional paradigm—state-led implementation of technological solutions—was consolidated despite the fact that the sheer diversity of social, political, and institutional contexts in the recipient countries should have made donors realize the need to contextualize technological choices to suit specific socio-political realities. As it was, however, questions about the efficacy of a centralized technological approach and its unintended consequences were not raised in any substantial manner at the societal level for almost two decades.

Until the early 1970s, questioning of the merits of this approach was muted: it was simply assumed the state should occupy a commanding role in the development of the economy and that technology had all the answers. It was during this period of blind faith in the 1950s and 1960s that Nepal began its development journey, one that cemented the Panchayat system's hegemony. The uncomfortable questioning of the basic tenets of the then current development paradigm was simply not allowed but blind faith in the existing state led growth paradigm did not see the majority of the Nepali population even receive basic services. Gradually, the fact that this conventional approach was not meeting its objectives in Nepal—or in almost all other developing countries, for that matter—became the central agenda of the 1972 Stockholm Conference on Environment and Development. Participants from South Asia emphasized three imperatives of foreign assistance: survival, social justice, and sustainability.

In the 1970s, citizens and organized groups began questioning the conventional paradigm through civic actions by women like the Chipko (Hug-the-Trees) Movement. These actions had a common thread: all were non-violent actions led by individuals who espoused Gandhian philosophy. Many actions questioned the results of the development of large dams. Even though the stored water helped generate electricity, provide irrigation, support human consumption, industrial use, and aquaculture it caused significant social displacement, loss of environmental resources, and pushed the poor and disadvantaged into further social and economic marginalization. In Nepal, debates began only in the early 1990s, after multiparty democracy had been re-established and the political space began to accept questions and critique. Though muted, alternative voices began to be expressed and inform the public dialogue.

Though it did not have many dam projects, Nepal nonetheless had its share of debate on projects such as Arun III and West Seti as well as about the Mahakali

Treaty and the saga of the bankrupt Enron Company (which was slated to invest billions of dollars in the developing the massive Karnali Chisapani Dam). Much of the criticism in Nepal focused on the fact that Nepal still lacked basic services such as electricity and that the proposed projects did nothing to provide services, create local employment, boost local economies, or encourage self-reliance. On the contrary, they perpetuated Nepal's dependence on external expertise.

The early 1990s was a watershed in global history as well as on water. The Berlin Wall had collapsed, the Soviet Union disintegrated, and the Internet and World Wide Web came into use, while globalization emerged as new phenomenon. In 1992 the Dublin Conference enunciating the four principles on water management emphasized that freshwater is a finite and vulnerable resource, advocated a participatory approach to water development and management, recognized women's central role in water provision, management, and protection, and posited water should be considered an economic good. The same year, the Earth Summit passed seven statements on water that addressed taking a holistic approach to water, assessment, protection of ecosystems, ensuring access to drinking water and sanitation, urban water, food production with emphasis on women, and addressing the impact of climate change.

During the 1990s, debates about Tehri High Dam and Sardar Sarovar Project in India and Bangladesh's Flood Action Plan were at their peak. The era also saw neoliberal ideology promoted as a solution for governance and with the idea that government has very little role in statecraft and that markets would do all the needful. It was again a period when activists found support in the communication revolution that was taking place. In this decade the dot-com bubble boomed and busted in the US but by the end of the century, internet-based communication was widespread. Just a few years later, social media had come to occupy central stage for the transfer of information as well as for the pursuit of political and social agendas including issues such as water and environmental management.

The 1990s was also an era of creative engagement among civic actors in South Asian countries[16] and beyond. Some of the finest South Asian minds,[17] seeking to combine the natural and social sciences, came together to challenge the prevailing paradigms of development. They debunked the notion that those who kept themselves on pedestals, best practiced development and questioned forced top-down decisions about water, development and societal concerns onto non-empowered and disenfranchised populations[18] wrote of the South Asian water landscape during that period: "It is high time to end the old fiction—started during the days of the empire, uncritically accepted by political leaders after independence and vigorously encouraged by those who benefit most from it—water development is technical

[16]Behera, N. and Rizvi, G, (1998). *Beyond the boundaries: A report on the state of non-official dialogues on peace, security and co-operation in South Asia.* Toronto and New York: Governing council of the University of Toronto and the Board of Governor of York University.

[17]Guha R. (2013). The past & present of Indian environmentalism. The Hindu.

[18]Bottral, A. (1992). Fits and misfits over time and space: Technologies and institutions of water development for South Asian Agriculture. *Contemporary South Asia,* 1 (2), 227–247.

subject best left to technical specialists to deal with—preferably behind closed doors."

Civic actors raised fundamental and critical questions that made donors and recipient governments equally uncomfortable: will the current development pathway widen the disparities that already exist between the haves and have-nots? Critics highlighted the constraints that were keeping large segments of the South Asian population poor and questioned why they did not have a decent education, dignified employment, safe housing or good health.

The formation of the World Commission on Dams (WCD) in 1998 was a crucial point in the debate on dams, water and development. The commission brought together both dam proponents and dam opponents to review the performance of dams across the world and to propose a new framework for building them. Two years in the making, its report, "Dams and Development," proposed a new framework and guidelines for developing dams[19] without making the mistakes of the past, but few governments or private-sector actors paid any attention. For the most part, it was only civil society actors who suggested that the principles recommended by the WCD be internalized. Their advocacy made serious effort at pushing the political landscape across the globe and in South Asia to make changes.

Unfortunately, with the beginning of the new millennium, the space for civil society voices that were active in the water debate gradually began to shrink as the *mantra* of economic growth began to take precedence over questions of environmental sustainability and equity. Despite the Wall Street crash of 2008, market-led growth continues to dominate the development discourse and state-sponsored environmental safeguards continue to be dismantled, disregarded, or even abused. The liberalization of the economy and the entry of market in the 1990s did bring about reforms in sectors such as aviation, banking, health care, and education, but the change was not entirely advantageous as political actors took over private business and business leaders assumed political roles. Social and environmental safeguards are now seen as bottlenecks that delay economic development and the national media badmouths families who seek compensation for land that the state acquired under the right of eminent domain, as "anti-development." The state and the market together seek to drown out the voices of caution.

The adverse outcomes of this change are clear. All over South Asia environmental quality continues to deteriorate. In all major cities, the level of air pollution is sky high and the rivers that run through them are effectively dead. Groundwater tables are alarmingly low and untreated solid waste is dumped on common land and wetlands and in rivers. Forest health continues to decline, and stone, sand, and other minerals are wantonly extracted to meet the needs of growing urban regions with no thought of the future.

At the same time, the threat of organized violence, often spurred by resource shortages like Darfur in Sudan, where herders and farmers compete, haunts

[19]WCD (2000). *Dams and development: A new framework for decision-making*. The report of the world commission on dams. London and Sterling: Earth scan Publications Ltd.

countries and societies, remain. Such violence isn't limited to suicide bombs in faraway places Nepal has its own share: the so-called People's War launched by the Maoist party in 1996 claimed 17,000 lives before it ended in 2006 and the entire royal family was massacred in 2001. These events paved the way for Nepal's being declared a secular federal democratic republic in 2006. How exactly this political change, coupled with the changes in geography and developmental paradigm alluded to above, will impact the future use and management of water and transboundary water governance is still unclear. This is so because parties that were not involved in the drafting of the constitution contest the constitutional provision of the demarcation of the federal boundary stipulated in Nepal's new constitution. At the time, India imposed an unofficial blockade of Nepal and relations between the neighbours reached a new low. This was inexplicable as Nepal and India share a similar culture and the citizens of each country can travel to the other country without any visa. No other countries have such a provision.

It is probable that investments in dam construction will increase for two reasons. First, the demand for energy will rise as India and China continue along their current development trajectory. Second, hydropower is being promoted as a low-carbon solution to climate change. Thus, it is likely that private companies and development banks will increase their financing of new dams as sources of clean energy. In addition, new storage reservoirs will be constructed to buffer shortfalls of water associated with climate change-spawned droughts.

However, climate change introduces two risks to such reservoirs. First, drought may lower river flow so much so that the designed storage capacity of a storage dam cannot be filled. Second, the likely increase in sediment production will deplete the storage capacity of dams faster than the design estimate predicts. Both risks need to be assessed before a decision to invest is made. At the same time, the social and environmental costs of the submergence of a large area behind a storage dam must be addressed institutionally and cannot be glossed over by a vague promise of a resettlement plan. The challenge in this emerging landscape is to ensure that voices that caution and raise concerns about equity, injustice, and marginalization are not pushed to sidelines or dismissed altogether as hindrances to growth.

5.5 Emerging Challenges

By investing in a transboundary river, nations aim to generate benefits such as electricity, irrigation, flood mitigation, and drinking water that can be shared by users on both sides of political boundaries. Theoretically, at least, outcomes are better if neighbours come together to harness a transboundary river than if they act alone. In practice, however, experiences in South Asia suggest that unless there are national policy reforms, the local level cannot avail intended benefits. In fact, interventions in transboundary rivers have actually resulted in unintended degradation of environments and hardship for local people. The allocation of benefits both among neighbouring countries and within a single nation has been debated, contested, and challenged. Official mechanisms such as bilateral joint river commissions flounder as they lack the mandate to address the emerging challenges of water, nature, and society, which include data-sharing, joint investments, equitable benefit-sharing, joint monitoring, joint operations and management, and joint problem-solving. To improve the present situation regarding water use and management and poverty in South Asia, we must undertake the following measures:

5.6 Discard the A-Hydrological Notion that Flowing River Water is Wasted

The conventional approach to water, which began in the post-industrialized era and was consolidated in the 20th century water development paradigm, is founded on the erroneous notion that environmental resources such as water are to be consumed or put to work only for the benefit of humans and that any water which flows into the sea unused is wasted. The Columbia Basin Development League of Western United States captured this belief in its assertion: "Every drop of water that runs to the sea without its full commercial return is a waste".[20] In the 1890s, the head of hydrological studies in the U.S. Geological Survey wrote, "The stability and permanence of the commonwealth are assured by the conservation of the water resources which otherwise would go to waste".[21] The notion of conservation was not protection but that of total control by the damming and storing of the natural flow, thereby diverting it for human uses. Many years later in 1920 when Franklin D. Roosevelt while campaigning for vice presidency saw the Columbia River, his first instinct was for control. "I could not help thinking of all that water running unchecked down to the sea", he said as soon as he got off his train at Portland.[22] As President of the U.S. many years later, through the New Deal, he played a key role in ushering an era of dam building in the U.S.

Closer to home, in the heartland of Awadh, in the early 1900 the government of the United Province pushed for developing the Sharada canal system. The government had argued that the people of Awadh did not use the Sharada River optimally and its water "went to waste".[23] The government proposed that the river water should be channelled via the Agra canal to Punjab for uses. In 1910, the government of the United Province and the Rana government in Nepal began

[20]Harden, B. (2012). *A River Lost, The Life and Death of the Columbia*, W. W. Norton and Company, New York, London.

[21]Worster, D.(1985). *Rivers of empire: Aridity and growth of the American West.* New York: Oxford University Press.

[22]Ibid.

[23]Whitcombe, E. (1983). Irrigation and railways. *The Cambridge economic history of India*, Kumar, D. (Ed.), 2. Cambridge.

negotiation on sharing the Sharada River at Banabasa and in 1920 concluded the Sharada Agreement. The Sharada barrage and canal system was completed in 1928.[24]

Thus, the 19th and 20th century capitalism and capitalist march progressed[25] with the backing of dominant knowledge brokers such as USGS and the agency of U.S. presidency conceiving flowing river unused as a waste. This conception of 'waste' ignores the fact that rivers within the hydrological system link oceans and landscape continuously, to which the idea of sink-and-source such as carbon cycle does not apply. Unfortunately, this notion of flowing river as a waste continues even today in developing countries.

Develop Tools and a Framework to Deal with the Likely Increase in the Incidence of Extreme Events

Many recent events have brought to the fore the limitations of conventional approaches in dealing effectively with the challenges of flooding in South Asia. One such event was the 2008 Koshi flood. Despite a standing treaty between Nepal and India on the Koshi River and regarding the maintenance of embankments, neglect resulted in a breach that seriously affected about 3.5 million people in Bihar and 50,000 in Nepal.[26] Ironically, the Koshi treaty contains no provisions for responding to such a catastrophe despite the fact that its primary objective is to mitigate flooding. Another event, the 2009 Mumbai flood, serves as a reminder of how and why cities are vulnerable to floods and the 2015 Chennai floods showed that even mega cities are vulnerable. In fact, cities are particularly vulnerable since they block natural drainage systems and the infrastructures they construct in their stead cannot handle the deluge resulting from extreme rainfall events. The 2010 Indus floods affected all aspects of life in Pakistan, an outcome that exposed the limitations of both state agencies and 'hard' measures such as embankments and dams for mitigation. Other extreme floods—the 2013 West Nepal and Uttarakhand floods, the 2014 Kashmir flood, and the 2014 West Nepal floods[27]—all saw the collapse of local livelihoods, loss of vital household food supplies, and damage to transport, communication, energy, health, and drinking water supply systems as well as the failure of the local institutions on which the local populations depended. These floods had immediate consequences for people across all levels of society, but the impacts on the poor and marginalized populations were most direct and severe. One of the reasons for the severity of the impacts was that, in all case, the

[24]Dixit, A., Adhikari, P. and Thapa, R. R. (2004). Nepal: Ground realities for Himalayan water management. *Disputes over the Ganga*. Kathmandu: Panos Institute of South Asia.

[25]Ibid.

[26]Dixit, A (2009). Koshi Embankment Breach in Nepal; Need for a Paradigm Shift in Responding to Floods *Economic and Political Weekly*.

[27]MacClune, K., Venkateswaran, K., Dixit, K. M. Yadav, S., Maharjan, R., and Dugar, S. (2014) Risk Nexus Urgent case for recovery: what we can learn from the August 2014 Karnali River floods in Nepal Zurich Insurance Group Ltd and ISET-International.

wetlands that function as a natural cushion for flooding had been severely encroached and natural drainage blocked. Clearly flood mitigation approaches need re-conceptualization.

5.7 Make Water Studies Inter-Disciplinary

The approach to transboundary water is dominated by the disciplines of international law, economics, and civil engineering, but their perspectives are limited. International law considers water in terms of its being in a river course within the territorial unit of a nation-state, but, while this is a necessary understanding of the nature of water, it is not sufficient for dealing effectively with emerging water challenges. An economist considers how water can be exchanged and traded between various places and users to foster growth. To promote transactions, engineers design and build hydraulic structures that modify the natural stock and flow of water, an approach that is the hallmark of the conventional water development and management paradigm. Unfortunately, this approach has tended to divide water issues along ideological lines, setting up oppositions between local and centralized approaches, supply-side and demand-side dynamics, and market efficiency- and -based regulations. Because such an approach lacks inter-disciplinarily, it relegates ecosystems, water quality, and social, institutional, and individual behaviour issues to the back burner.

This approach is particularly damaging in the South Asian context, where social inequality, disputes, and asymmetrical power relations among nations are rife. The preference for centralized, technologically guided, and market efficiency goals and supply-side solutions has exacerbated scarcity and increased the number of people who have lost access to clean water and livelihood sources. When access to clean water is limited, health, basic needs, and poverty immediately become a concern. In the urban and peri-urban slums of South Asian cities, poor families often pay a substantial portion of their income in order to purchase enough water simply to meet their basic needs. Given the high rates of population growth in cities and the prevailing demographic shift from rural to urban areas, water stress in GRB countries is bound to increase and will only be compounded by climate change uncertainties.

5.8 Nurture Plural Perspectives

Current water challenges are much more complex than the simplistic "small versus big" debate of the past. While social movements in the later decades of 20th century highlighted issues like ecological degradation, inadequate rehabilitation of project-displaced populations, and social injustice, a set of second-generation concerns has since emerged. These, too, must be addressed even as persistent first-generation problems demand attention. These second-generation questions stem from two new key drivers of dam development, the changing political landscape and climate change. Climate change means that the hydrology of the future will no longer be that of the past and raises the possibility that the methodology behind the design and economic evaluation of proposed dams might be so outdated as to be ineffective or even damaging. The information revolution has brought political awareness even among the most marginalized of communities, enabling them to find new ways of engaging in decisions over a variety of issues, including water management. Responding effectively to these challenges requires having access to creative modes of engagement at all levels. Each mode, in turn, must include in-built compliance and accountability while the political space allows plural voices and perspectives to be expressed.

5.9 Institutionalize Multi-stakeholder Dialogue Platforms

The 1990s saw the region's non-state actors take up initiatives to promote dialogue about transboundary water issues. To a limited extent, these initiatives helped open informal channels of dialogue that explored alternatives without requiring the commitment of national governments. The outputs of these initiatives, however, remain fragmented; they have not been synthesized to generate coherent lessons for governments wishing to devise new policies. The findings of the research conducted under these initiatives have not been communicated to the public either. In recent times, the private sector has been a major player in the water market, in supplying water in tankers, in agro- and food system businesses, and in hydropower generation. Because public- and private-sector actors dominate the policy space, the issues of water quality, marginalization, and injustice have gradually become of less interest even as these challenges have become more serious. The governments of GRB countries, unlike many industrialized countries, have not internalized lessons from past efforts and no process of social learning has taken place. There is a need for guidance groups and platforms that foster mutual and shared learning among disciplines, sectors, analysts, and common people.

5.10 Conclusion

How can the opportunities that rivers offer be harnessed to our benefit without compromising the inherent integrity of water and the ecosystems and societies dependent on it? To answer this important question we still need to ask more questions: How do we accomplish this goal without compromising relationships among riparian countries? How do we improve the efficiency and productivity of water use per drop and, in doing so, contribute to food security, poverty reduction, and economic development? What policies would bring about desirable changes?

How do we ensure that equity and distributive justice is not compromised in the name of development? Should policies remain national or do we need regional policies? Because water in the GRB and, for that matter, South Asia as a whole already faces fundamental crises of efficiency, ecological sustainability, and equity, answers are needed quickly if we are to ward off a catastrophe.

Responding to water challenges will require us to examine the notion of governance by asking ourselves how and by whom water management is governed and under what conditions different approaches to water uses and management should be selected.[28] The following questions also need answering—who has a seat at the table? How are the impacts of water variability distributed? What types of information are collected for decision-making and who has access to them? What incentives do individuals and groups face in choosing an option and how do these incentives shape their perceptions of and responses to emerging water problems? Who takes part in the conversation? These questions are rarely asked in the context of transboundary water governance. The atmosphere of mistrust prevailing among countries has pushed such questions under the carpet, thereby exacerbating the abuse and degradation of water commons and making more and more families vulnerable to various change drivers, including climate change.

Dealing with the complex contexts both within any given country and at the transboundary level requires a new approach. Overcoming the challenges outlined above is not simply about two governments coming together to build a project. The real challenge is framing policies that will help overcome those and any future constraints that might emerge. It lies in charting new pathways for bringing about transformative changes and will require internalizing the meaning of water in the real sense.

Water in the GRB and in South Asia as a whole faces myriad challenges. Glacier recession due to climate change and other change drivers will be a serious blow for lowland water use and management, as users will have to grapple with the consequences of the upland loss as well as with abused water as well as climate change impacts. Both upland losses and lowland abuses of water are outcomes of a context characterized by increasing uncertainty and variability as well as rapid changes, misappropriation, and governance deficit. Conventional methods cannot address these problems because they are the outcome of the context, which created them in the first place. As Albert Einstein has reminded us, "We cannot solve our problems with the same thinking we used when we created them." We do not yet know what these new approaches need to be or the pathways we need to travel to get to them, but the need for change is abundantly clear. The present is the best time to begin a journey that will recognize water for what it is—an element essential for nature, societies, and ourselves—and take into account that perspectives on how water should be used and managed will inevitably differ.

[28]Moench, M., Dixit, A., Janakarajan, S., Rathore, M. S. and Mudrakartha, S., (2003) The Fluid Mosaic: Water Governance in the Context of Variability, Uncertainty and Change: ISET and NWCF.

We must, through creative engagement, share these various perspectives in all their diversity at many levels, including the communal, regional, national, and international. The purpose of such engagement is not only to enable government, civic, and private sector actors to choose a set of technically viable solutions to water problems that can be integrated in order to meet multiple demands. Both the technical solutions and the sort of integration required depend on the context and the specific nature of the water problem being addressed.[29] Engagement is necessary to identify and integrate different forms of knowledge from diverse sources as they emerge from the wider socio-political processes. Our challenge is to make governance, particularly the foundations on which day-to-day decisions and courses of actions rest, accountable.

Almost two decades ago Imtiaz Ahmed, a political scientist from Bangladesh, Ashis Nandy, a social psychologist from India, and I, a water resource analyst from Nepal, wrote in *South Asian Manifesto on Water, Power, and People*:

> We need to integrate the global (macro) and the local (micro), accommodate a plurality of views, generate options reflecting the larger reality of water in nature and human society. This is not possible without altering structure and nature of current decision-making approach, and the educational context within which they are generated.[30]

A different educational context and a different decision-making approach are missing in the transboundary water policy terrain in the GRB and in South Asia as a whole. Creating these two things could be a starting point for conceiving of a new compact for stewarding South Asian water. As a means of economic production and source of drinking water, rivers will continue to be used by humans directly and indirectly, but as Donald Worster (1985) has reminded us, “a river is also an entity unto itself, with its own processes, dynamics, and values. It is a sacred being, something we have not created, and therefore worthy of our respect and understanding”. South Asians indeed consider rivers ‘sacred’ and must imbibe this ethos to begin a new water journey with alacrity. Rivers need more respect and much deeper understanding than at present.

---

[29]Ibid.

[30]Imtiaz A., Dixit, A., and Nandy, A. (1997). *South Asian manifesto on water, power and people*: RCSS Colombo.

About the University of Dhaka

On the first day of July 1921 the University of Dhaka opened its doors to students. The University was set up in a picturesque part of the city known as Ramna on 600 acres of land.

The main purpose of the University was to create new areas of knowledge and disseminate this knowledge to the society through its students. Since its inception the University has a distinct character of having distinguished scholars as faculties who have enriched the global pool of knowledge by making notable contributions in the fields of teaching and research.

A new phase began in the history of the University with the emergence of the People's Republic of Bangladesh in 1971. This has been a phase of development, expansion and consolidation of earlier gains. The University has assumed a central role in the academic pursuits of the region including this new nation. The University of Dhaka has passed through tumultuous times at different periods of the national history and played vital, at times pioneering, roles in all critical junctures in the making of this great nation. The University played a central role in the Language Movement of 1952 that ultimately culminated in the recognition of Bangla as the State Language. Scores of students, teachers and employees of the University of Dhaka laid down their lives for the independence of the country. Just after the creation of Bangladesh, the government proclaimed the University of Dhaka Order 1973 whereby democratic norms and autonomy became integral features of the institution.

The University of Dhaka is dedicated to the advancement of learning, and is committed to promoting research in all fields of knowledge. As there are plans for further expansion of facilities, plans for new avenues and opportunities, the course curricula are updated and new research projects are undertaken every year. As the pioneer and the largest seat of learning in the country, the University of Dhaka has taken the task to foster the transformation processes of the individual students and the country as a whole through its educational and research facilities keeping up

I. Ahmed (ed.), *South Asian Rivers*, The Anthropocene: Politik—Economics—Society—Science 21,
https://doi.org/10.1007/978-3-319-67374-5

with demands of the day. The University of Dhaka is at this moment one of the leading institutions of higher education in Asia.

The University of Dhaka is well prepared to meet the challenges of the future days with its spirit of freedom, justice and truth as a foundation concomitant to the objectives envisaged by the founding fathers.

For further details please visit website: www.du.ac.bd

About the Bangladesh Environmental Lawyers Association

Bangladesh Environmental Lawyers Association (Bengali: বেলা), or BELA (ব় লা), is a non-profit legal organization established in 1992 to assist efforts to protect the environment.

BELA started its mission as an advocacy group of young lawyers developing legal techniques and strategies to protect the environment and defend rights of communities dependent on nature and natural resources. Founded by Dr. Mohiuddin Farooque, the organization has paved the way for environmental legal activism in the country and has been the pioneer of public interest litigation in Bangladesh. BELA has adopted various means to create awareness amongst the major actors and the common people about their environmental rights and duties. BELA has published books and other awareness materials, all having relevance to the concept of environmental protection and justice. It is also active in promoting people's right to information and the concept of participatory law-making in the country. BELA is a member of Friends of the Earth International, IUCN-the World Conservation Union, Environmental Law Alliance Worldwide (E-LAW), the NGO Platform on Ship-breaking and the South Asian Watch on Trade, Economics and Environment (SAWTEE). In 2003, BELA has received the Global 500 Role of Honors of the United Nations for its contribution to the promotion of environmental justice. In 2008, BELA was one of first four recipients of the Environment Award given by the Department of Environment of the Government of Bangladesh. The Chief Executive of BELA has received the Goldman Environment Award (2009), known as the Noble Prize in the environment sector for her successful legal battles in regulating deadly ship breaking industry in Bangladesh. She has also received the Ramon Magsaysay Award (2012) which is widely respected as the Noble Prize for Asia for her uncompromising courage and impassioned leadership in a campaign of

I. Ahmed (ed.), *South Asian Rivers*, The Anthropocene: Politik—Economics—Society—Science 21,
https://doi.org/10.1007/978-3-319-67374-5

judicial activism in Bangladesh that affirms the people's right to a good environment as nothing less than their right to dignity and life.

Address: Bangladesh Environmental Lawyers Association (BELA), House-15A (4th Floor), Road-3, Dhanmondi R/A, Dhaka-1205, Phone: +88-02-58614283, 58610311, Fax: +88-02-58612957.
Email: bela@bangla.net
Website: belabangla.org

About the Editor

Imtiaz Ahmed, Professor of International Relations and Director, Centre for Genocide Studies at the University of Dhaka, Bangladesh. Professor Ahmed was educated at the University of Dhaka, Carleton University, Ottawa, and the Australian National University, Canberra. He is also currently Visiting Professor at the Sagesse University, Beirut, Lebanon. His research interests include South Asian politics, religion and culture, tolerance and terrorism, the politics of development, refugees and migration, and women and the environment. Professor Ahmed is the recipient of various awards and honours. He has been a fellow in the following institutions: Ford Foundation Fellow at the University of Oxford; Asia Fellow at the Centre for the Study of Developing Societies (CSDS), Delhi; Rockefeller Fellow at the Rockefeller Foundation; Japan Foundation Fellow at the Yokohama City University; Research Fellow at the Institute of Southeast Asian Studies, Singapore; and Foreign Policy Fellow at the University of Maryland and College Park.

He has authored, co-authored, or edited 18 books and 6 monographs. More than 110 research papers and scholarly articles have been published in leading journals and chapters in edited volumes. His publications include *Contemporarising Tagore & the World* (2013), *Women & Militancy: South Asian Complexities* (2011), and *Terrorism in the 21st Century: Perspectives from Bangladesh* (2009) and his recent publication is an edited volume titled: *Human Rights in Bangladesh: Past, Present & Futures* (Dhaka: University Press Limited, 2014). His forthcoming publication is *People of Many Rivers: Tales from the Riverbanks* (Dhaka: University Press Limited, i.p.). His CV can be viewed here at: https://rcss.org/wp-content/uploads/2016/01/Prof-Imtiaz-Ahmed-CV.pdf.

I. Ahmed (ed.), *South Asian Rivers*, The Anthropocene: Politik—Economics—Society—Science 21,
https://doi.org/10.1007/978-3-319-67374-5

Address: Prof. Dr. Imtiaz Ahmed, Department of International Relations, 6th Floor, Social Science Faculty Building, University of Dhaka, Ramna, Dhaka PO 1000, Bangladesh.
Email: imtiazalter@gmail.com
Website: https://rcss.org/about_us/imtiaz-ahmed/

About the Contributors

Ajaya Dixit (Nepal) is the editor of *Water Nepal*, a journal addressing Himalayan water and development; and Chairman of the Board of Directors, Nepal Water for Health (NEWAH), a Nepali NGO that has built water supply schemes serving over 350,000 people in Nepal. Mr. Dixit taught hydraulics at Tribhuvan University's Institute of Engineering until 1989 and recently authored the current leading text-book in hydraulics, published in 2002, Basic Water Science. He has worked extensively as a consultant on water resources and environment for bilateral and multilateral organizations in Nepal. Mr. Dixit served as a member of the National Preparatory Committee for the United Nations Conference on Environment and Development, Rio Conference (1992). He also represented the private sector in His Majesty's Government of Nepal Water and Energy Commission from 1994 to 1997.

Address: Mr. Ajaya Dixit, 48 North Street St. 7, Katmundu, CO 80304, Nepal.
Email: adbaluwatar@ntc.net.np
Website: http://newah.org.np/

Dr. V.G. Hegde (India) is currently working as an Associate Professor at the Faculty of Legal Studies, South Asian University (SAU). Before joining SAU he was an Associate Professor at the Centre for International Legal Studies (CILS), School of International Studies, Jawaharlal Nehru University (JNU), New Delhi. Dr. Hegde previously held the position of Legal Officer at the Asian African Legal Consultative Committee (AALCC), now Asian African Legal Consultative Organization (AALCO). He was part of the Indian negotiating team on several bilateral investment and Trade Agreements. He was a member of the Working Group on Arbitration and E-Commerce constituted under the auspices of the United Nations Commission on International Trade Law (UNICTRAL).

Address: Dr. Venkatachala G. Hegde, Associate Professor, Faculty of Legal Studies, South Asian University Akbar Bhawan, Chanakyapuri New Delhi–110021, India.
Email: vghegde@sau.ac.in
Website: http://www.sau.int/faculty/faculty-profile.html?staff_id=63

I. Ahmed (ed.), *South Asian Rivers*, The Anthropocene: Politik—Economics—Society—Science 21,
https://doi.org/10.1007/978-3-319-67374-5

Shafqat Kakakhel (Pakistan) has been a member of the SDPI Board of Governors since 2009. He is a former member of Pakistan's Diplomatic Service and also served as Deputy Executive Director of the UN Environment Programme for nearly a decade. During his tenure in Delhi, Shafqat Kakakhel represented Pakistan as Permanent Representative to the UN Environment Programme and the UN Centre for Human Settlements-HABITAT from September 1994 to August 1998. He was elected President of the UNEP Governing Council in May 1995 and was the Chief Negotiator for the G-77 in the preparatory meetings of the second World Summit on Human Settlements in 1995–6 and was elected chair of the main negotiations committee at the Summit held in Istanbul in June 1996.

Address: Amb. Shafqat Kakakhel, Sustainable Development Policy Institute (SDPI), Taimur Chambers Plot # 10-D (WEST), Fazl-ul-haq Road, Blue Area, Islamabad, Pakistan; and House # 15, Street # 83, G6/4, Islamabad, Pakistan.
Email: shafqatkakakhel@gmail.com
Website: http://www.sdpi.org/media/media_details1199-press-2016.html

Dr. Sudhirendar Sharma (India) is currently working as the Director, The Ecological Foundation, New Delhi, India. Sharma has a doctorate in environmental science from Delhi-based Jawaharlal Nehru University. He has experience of designing and implementing rural development projects at the grassroots; has been senior journalist with leading publications like India Today (weekly) and The Pioneer (daily newspaper); and has worked at the World Bank.

Address: Dr. Sudhirendar Sharma, 7 Triveni, A6 Paschim Vihar, New Delhi 110063, India.
Email: sudhirendarsharma@gmail.com
Website: https://www.linkedin.com/in/sudhirendar-sharma-72504524/?locale=de_DE and http://researchandmedia.ning.com/profile/SudhirendarSharma

Index

I. Ahmed (ed.), *South Asian Rivers*, The Anthropocene: Politik—Economics—Society—Science 21,
https://doi.org/10.1007/978-3-319-67374-5

MIX
Papier aus verantwortungsvollen Quellen
Paper from responsible sources
FSC® C105338

If you have any concerns about our products,
you can contact us on
ProductSafety@springernature.com

In case Publisher is established outside the EU,
the EU authorized representative is:
Springer Nature Customer Service Center GmbH
Europaplatz 3, 69115 Heidelberg, Germany

Printed by Libri Plureos GmbH
in Hamburg, Germany